HANDBOOK OF

survey
notekeeping

HANDBOOK OF
survey
notekeeping

F. WILLIAM PAFFORD

licensed land surveyor

JOHN WILEY AND SONS, INC.

new york   london   Sydney

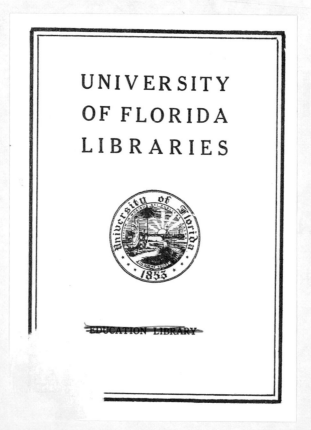

ISBN 0 471 65751 4
Library of Congress Catalog Card Number: 62–17467
Printed in the United States of America

To the memory of T. O. Molloy,

who, throughout his career,

constantly championed adequate notekeeping,

this handbook is respectfully dedicated.

sponsor's preface

Educational needs, regulatory situations, economics of surveying practice, apprenticeship programs, and other factors have made civil engineers and land surveyors realize the need for guides whereby their employees and neophyte surveyors can be directed toward acceptable standards of practice. This book furnishes such a guide for one aspect of surveying practice.

F. William Pafford has prepared this book as an educational project of the California Council of Civil Engineers and Land Surveyors. Although the Council has not edited the book, the material was prepared at its request and under its auspices.

The Council is composed of fourteen California associations of private-practicing civil engineers and land surveyors. There are two hundred seventy-five firms in the Council, with four hundred fifteen principals employing thirty-six hundred engineers, surveyors, technicians, and clerks. It is the hope of the Council that private practitioners beyond California's borders, and governmental agencies everywhere, will join the Council in a program to raise standards of performance in surveying through such means as are provided by this book.

THE EXECUTIVE COMMITTEE

May 1962

preface

The aim of this handbook is to produce, in a single volume, a complete, comprehensive coverage of the subject of survey notekeeping and, through the use of examples, not only illustrate the fundamentals of notekeeping but supply a ready reference for the man in the field to use as a guide when selecting the arrangement, form, and format of notes for varying types of survey work.

It is an impossible task to attempt to develop a set of standard forms covering all survey operations that would be acceptable to all surveyors because, even though certain operations in surveying may lend themselves to standardization, surveying in general does not. Furthermore, the survey situations under which the individual party chief is required to exercise personal judgment are so varied that to attempt to stereotype the presentation, that is, the field notes, would impair the development of the surveyor's most important attribute, *judgment*.

In the words "standard survey notes," the adjective "standard" does not mean meeting an exact rule, model, or form but is used in a broader sense. It means that survey information recorded within certain accepted limits of form, arrangement, lettering, neatness, etc., will have a strong similarity to other survey notes so prepared. When this similarity exists, the survey notes are said to be "standard."

The information contained herein is addressed to the party chief. Even though the recorder, or notekeeper, of traditional surveying still exists on certain types of survey parties, modern practice, especially in private work, places the responsibility for field notekeeping on the party chief.

It is assumed that anyone using this book is well based in the theory, required accuracy, and practice of surveying; no attempt is made to cover these subjects, although I have taken the opportunity to touch on certain survey fundamentals which are frequently neglected.

There are in use today a variety of sizes and types of field books, printed forms, etc., and there is considerable disagreement over the relative merits of loose-leaf books, transparent paper, duplicating carbon books, clip boards, sketch boards, and myriad notekeeping systems designed to fulfill the requirements of a particular firm's method of operation and type of work. Since it is the purpose of this book to develop a complete understanding of notekeeping for various types of surveying, the bound type of level, transit, and cross-section book without printed column headings has been chosen for the illustrations.

I do not intend to imply that printed forms should not be used; as a matter of fact, they are excellent for certain types of notes. I do intend, however, to discourage the reliance on printed forms to the exclusion of being able to do an adequate job without them. When the information in this manual is understood, the reader will be able to develop notes to suit any survey problem and will also be able to adapt himself easily to any forms or systems he may be required to use.

The handbook is divided into seven parts.

Part 1 is an introduction to notekeeping to acquaint the reader with the general requirements of field notes and to develop a philosophy that will allow a full appreciation of the information contained in the other parts.

Part 2 covers the arrangement and indexing of field books. The importance of this cannot be overemphasized, for often the user is confounded in attempting to follow the continuity of the work.

Part 3 explains abbreviations, symbols, and codes and their use and effectiveness.

Part 4 is devoted to the importance and arrangement of monument descriptions and associates the descriptions with types of surveys.

Part 5 covers the recording of support data and its effect on the confidence level of the survey.

Part 6 is designed to cover notekeeping on a variety of surveys and illustrates how notes may be prepared to fit any survey conditions.

Part 7 summarizes the subject of notekeeping and touches on the effects of future technological advancements.

The sample notes contained herein are full-size reproductions of field books and were prepared in pencil with the degree of neatness that would be expected of notes prepared under normal field conditions. The purpose is to supply examples with which the reader can compare his own work. For convenience, these notes are grouped at the back of the book and have a separate index.

The Appendix is a list of accepted abbreviations used in surveying.

Whenever the text refers to a figure, the reader is advised to make a detailed study of the sample notes referred to. In my experience, the best way to expand your knowledge of notekeeping is by studying notes prepared by others.

I am indebted to members of the California Council of Civil Engineers and Land Surveyors for their suggestions, advice, and assistance. Without their encouragement this book would never have materialized.

F. WILLIAM PAFFORD

Los Angeles, California
May 1962

contents

PART 1: introduction to notekeeping

1-1 DEFINITION OF FIELD NOTES

Field notes are a *written record*, arranged in a manner peculiar to surveying, showing *pertinent information, measurements,* and *observations* made by you in the field during the course of a survey, to be used and interpreted by a person having some knowledge of surveying.

The part of this definition that usually causes difficulty is the statement that you must keep your notes so that they can be "used and interpreted by a person having some knowledge of surveying"; that is, someone other than yourself, a person who may have only a limited knowledge of surveying.

If you will make the following hypothetical assumptions, whether they are valid or not, and bear them in mind whenever you are keeping notes, you will have laid the ground rules for notekeeping.

Assume that the person using your notes

(*a*) Has poor eyesight.

(*b*) Is not clairvoyant.

(*c*) Will try to place the blame on the field notes, if anything is wrong.

If this is the assumed situation, to overcome it and to protect yourself your notes must be

(*a*) Neat, legible, and clear.

(*b*) Complete and self-explanatory.

(*c*) Honest and self-checking.

Listed under Section 1-3 are some general points that must be observed if you are to keep good notes.

1-2 IMPORTANCE OF FIELD NOTES

Of all the operations accomplished by a survey party, the most *important* by far is *notekeeping.* It is obvious that no matter how carefully field work is performed or how expert the party may be, all is rendered valueless if the field notes are not intelligible to others.

Frequently surveyors believe that they have done an adequate job of notekeeping if the field notes, reinforced by their memories, are sufficiently comprehensive to allow them to be used for an immediate purpose. Obviously, this is not true. Field notes must stand by themselves and be interpreted without recourse to questioning the person who prepared them.

Incomplete and unclear notes result in the following.

(*a*) Lost time and additional costs for trying to decipher them.

(*b*) Necessity of returning to the job to clarify the notes.

(*c*) Erroneous information being placed on maps which can result in costly design errors.

(*d*) Inability to defend the work to others, especially in court.

(*e*) General mistrust of the party's work.

Actually, there is nothing difficult about keeping good notes if you understand how these notes are to be used and exercise the same amount of care and judgment in their preparation as you do in other survey procedures.

1-3 GENERAL REQUIREMENTS OF GOOD NOTES

Neatness

(*a*) Use a sharp pencil.

(*b*) Make liberal use of enlarged details; avoid crowding descriptions or sketches.

(*c*) Keep lettering parallel with or at right angles to the feature it refers to.

(*d*) Before starting a sketch, decide what it is to show and organize it. Do not start in one corner of the page and let it grow. Often you will have a map or plat that is to scale; use it as a guide in selecting the size you want. Sketches should be drawn to approximate scale.

(e) If you start a sketch that is not the correct size, void it and start over. Do not try to make the best of a bad start.

(f) Do not lay the book where it will become dirty. Keep your hands clean.

(g) Keep tabulated figures inside column rulings, not in the border. Keep digits and decimal points in vertical line.

(h) Use tools as shown in Figure 1 for drawing sketches.

(i) Do not overlap figures on each other or on lines of sketches.

Legibility

(a) Use Reinhardt's style of lettering for legibility and speed. (See Figures 2A and 2B.)

(b) Use pencil of at least 3-H hardness and press down so indentation is made in paper, but do not use a pencil so hard that notes are only readable in bright sunlight.

(c) Use symbols and codes to keep notes compact.

(d) Whenever possible, place North at top or left, and arrange sketch to read from bottom and right side of page.

Clarity

(a) Whenever possible start each day's work on a new page.

(b) Plan ahead and select a note form that is appropriate for the particular survey.

(c) Vary line weight when appropriate. Very often the use of colored pencils increases clarity, although colored pencils should not be used if the notes are to be photostated.

(d) Vary size of lettering for emphasis.

(e) Arrowheads and leader lines should be light and sharp and held to a minimum.

(f) Do not make ambiguous statements.

(g) Use proper nomenclature.

(h) Place a zero in front of numbers less than one. For example, 0.51 instead of .51.

(i) Exaggerate details on sketches for clarity.

(j) Line up descriptions and sketches so that it is clear what the descriptions refer to.

(k) Show a North point on all sketches.

(l) Be consistent in the way you record data. For example, in one place do not record the curb

elevation above the gutter elevation and on subsequent shots reverse the procedure.

Completeness

(a) Show all pertinent measurements and observations.

(b) Record complete data. For example, "Mr. John M. Jones, Job Superintendent, instructed me . . . ," not just "Superintendent instructed me. . . ."

(c) Review your notes before leaving job site to be sure they are complete.

(d) If you are in doubt about the need of recording any information, it is safer to record it.

(e) Do not leave anything to be assumed.

(f) If you intentionally leave something out, say so. For example, "Fence lines along north and west property lines not shown."

Self-explanatory

(a) Make maximum use of explanatory notes.

(b) Place title at beginning of each type of work stating what the work is.

(c) Show closing data with appropriate arithmetic checks and closures.

(d) Record sufficient data so field computations can be checked later in the office.

(e) Make your statements positive.

(f) When a point is established by record, by prorate, or by intersection, etc., state this fact. For example, "Set lead and tack by proportional measurement."

(g) Always cross-reference. When work is continued on another page, note this fact. Do not leave it to someone to guess.

Honesty

(a) Record exactly what you did in the field at the time you did it, not later from memory.

(b) Do not record measurements made by others unless you note this fact.

(c) Numbers should show degree of precision. For example, rod readings taken to the nearest 0.1 foot should be recorded 6.3, not 6.30.

(d) If you have measured a distance to the nearest 0.01 foot but question the accuracy, note this. For example, $321.47' \pm 0.10'$.

(e) Erasures are not permitted in field books. If an item is recorded in error, draw a line through it without destroying its legibility and record the correct item above.

(f) Avoid copying notes. If it is absolutely necessary, the part copied must be so marked and the original submitted as a part of the notes. For example, if the chaining notes are kept in a peg book and the totals are copied into the field book, the totals must be marked "copied from peg notes" and the peg book must be submitted as a part of the notes.

(g) When adding pickup data to notes previously prepared, date and initial each new entry.

(h) Record actual measurements, not what they are supposed to be.

(i) Never, under any conditions, falsify notes. This is fraudulent and dishonest.

Self-checking

(a) Field work and notes should be planned and kept in such a manner that the work can be checked without a return visit to the field; the office man should be able to take the field notes and, by calculation, prove the work to be correct. For example, in leveling, if you start and end on the same bench mark, the office man can check that all your turning point rod readings were read correctly, but he will not know whether the bench mark you used was the one you described; there may be two spikes in the same telephone pole. Had you started at one bench mark and ended on a different one, the office could positively prove your work.

(b) In boundary work or on building layout show the closing angles and distance you measured, not what they are supposed to be. Obviously, every distance you measure or rod you read cannot be checked, but the main scheme should be done in a way that is self-checking. It is a matter of judgment what part of the survey warrants checking. For example, to locate trees by one angle and one distance may be sufficient, but if you feel that their locations are important, perhaps you should, in addition, draw a sketch showing the distance between trees with the end trees tied to the property line. This will allow the office to check the correctness of the locations of all the trees.

(c) It is of the utmost importance that, on every survey, check measurements be taken to prove the main scheme.

1-4 GENERAL COMMENTS ON NOTEKEEPING

As you study the sample notes in this handbook, you will observe that all the notes, regardless of the type of survey being illustrated or the form and format of the page, have seven things in common. They are neat, legible, clear, complete, self-explanatory, honest, and self-checking. These are the attributes surveyors have in mind when they say "These are good, standard notes."

There is an additional ingredient necessary in notekeeping. That is pride. You should always strive to turn in a set of notes you can be proud of. Your field notes, more than anything else you do, create an impression on others of your ability and integrity.

Regarding supplemental information pertinent to the survey, do not hesitate to put written explanatory notes in the book, even if it takes several pages. For example, if you are run off the property by an irate owner, say so in your notes. If you are told by a neighbor he saw someone move a stake, note the neighbor's name and the pertinent part of the conversation. Remember, the notes are a record of *pertinent information, measurements,* and *observations.* If you observe something you feel is pertinent evidence, put this in your notes; it may win a law suit for your firm. Often, on a large survey which lasts several months, it is advantageous to keep a daily diary in a separate field book. It is permissible to cross-reference to a diary if it is submitted as part of the field notes.

A source of much difficulty for a party chief is to be given incomplete instructions about the type and amount of information that is to be shown in the notes. For example, the job work order may read "Complete boundary survey and topographic map for service station design," and may fail to mention that driveway easements are needed and the topography should be extended in the parkway 50 feet beyond the usual site limits.

It is not the party chief's responsibility to second-guess the person giving job instructions, but it is his responsibility to be sure that he has received clear instructions. Never start a job until you know positively what it is you are supposed to do.

A party chief should request that all instructions be given him in writing, and then be prepared to ask questions, based on past experience, that will define clearly the extent of the work. In the example stated, the party chief should have asked specifically "Do you require any offsite topography?"

Unfortunately, people ordering field work too often assume that the party chief is a mind reader, and that he will obtain the exact amount of detail to the proper degree of accuracy when he has been given only sketchy instructions. Often a party chief attempts to overcome the imprecision of his instructions by locating everything in sight with too high a degree of accuracy. This is a false precaution. You will be criticized just as severely for obtaining too much data as for obtaining too little data and for spending extra time obtaining an unwarranted degree of precision.

At this point, the whole subject of notekeeping may appear to be an insurmountable problem, but the reader should not be discouraged. An experienced party chief, given adequate instructions and exercising good judgment, can obtain the proper amount of detail to the proper degree of accuracy and, by a carefully planned field note form, can present the information so only the correct interpretation can be made.

PART 2: arranging and indexing the field book

2-1 SETTING UP A FIELD BOOK

When a party chief receives a new, blank field book, he is obligated to do the following before entering any survey notes.

(*a*) Note the company name, address, and phone number, if they are not already stamped in the book. If the book is lost, the finder will then know whom to contact to return it. The back of the front cover is an appropriate place for this information, and it is preferable to use ink. This information must be placed inside the book even though it may also appear on the outside. Information on the outside cover of field books often becomes illegible from handling. This information should appear in the following form:

> Survey Book S-566
> If lost, will finder please contact
> A. B. Smith & Associates, Surveyors
> 1234 Mill Street
> Los Angeles 57, California
> Phone: DUnkirk 1-2345

If your company will pay a reward for the return of the book, this should also be noted.

(*b*) All pages in a book must be numbered before field work begins. When the book is open, you see two pages, one to the left and one to the right of the book binding. Only the right-hand pages are numbered. The number is placed in the upper right-hand corner of the right-hand page. Every pair of pages is so numbered. Thus page 1 has ruled lines on the right-hand side only, the left-hand side being a blank fly leaf which is not used. Also, the last ruled page in the book is a left-hand page opposite either a fly leaf or tables, and this page is not used.

Once the pages are numbered, no page can be removed without upsetting the numbering system.

(*c*) The first two or three pages are headed "index" and are reserved for future indexing.

Comments. Do not use the fly leaves or the insides of the book covers for notes, calculations, additions, etc. To do so immediately raises a question to the person using the notes: "What is this, and why isn't it on the appropriate page in the notes?" Leaving these pages blank and clean also contributes to the general neatness of the book.

2-2 INDEXING

General

It is the responsibility of the party chief to index his work upon completion of each job or each appropriate increment.

An index is a time-saving device to assist the person using the book to locate quickly what he is looking for. Therefore, to index each page defeats the purpose of indexing, for it would be almost as quick to search through the book as it would be to search through an index listing every page.

The index is arranged to identify the work briefly by name or description and to refer to the pages where the work may be found.

If the book is properly subdivided by title pages, as later explained in Section 2-3, the index need only condense the job designation and work description with the proper page number reference. This will alert the user to which pages he is looking for, and when he reads the title page, he will know whether this is the work in which he is interested.

Indexes are usually arranged chronologically; as each piece of work is completed, it is indexed. The book may not be used again in the field for several weeks, but, when it is, the next completed increment of work is then indexed.

The arrangement of the index and the information required to be shown in the index vary somewhat, depending on the type and volume of work.

When a book is used for a number of small, unrelated jobs, the index can be very simple, showing only the job designation, type of work done, and page reference.

On larger jobs, a complete book or a series of books may be assigned to be used exclusively for that job; then it is often convenient to split the work and the index, keeping certain operations separate. Or it may be that all of one survey operation is kept in one book, in which case more detailed indexing is required.

On surveys of large magnitude, requiring a number of field books, many methods are used for master indexing. Most firms also have a master index system covering all the firm's work. Master indexing is considered an office procedure and is not covered here, except to note that the existence of a master index does not relieve the party chief of the responsibility of indexing within the field book.

Chronological indexes

Figure 3 gives an example of simple chronological indexing. Note that for each job indexed, the job designation, type of work done, page reference, and job number are given.

Under Item 5, which is the fifth piece of work indexed, the type of work "topographic survey" is subdivided into three classifications, "control traverse," "control bench levels," and "transit topo," each having its proper page reference as a further aid to the user in locating the work.

Note that the subheadings are emphasized by indenting.

Figure 4 is an example of chronological indexing with the added cross reference to a diagram "(See diagram for index of stations)," thus further subdividing the index to show page references for specific stations occupied.

Combined index and diagram

Figure 5 is an example of combining a triangulation diagram and an index, illustrating the page referred to as "2Rt," shown in Figure 4.

With this combination, care must be exercised to insure that the indexing does not impair the clarity of the triangulation diagram.

The use of a combined index and diagram does not eliminate indexing the work on the regular index page with the proper cross reference; therefore the diagram may be placed at any appropriate location in the book and does not have to be in the front portion.

Graphic indexes

The purpose of a graphical index, shown in Figure 6, is to allow the person using the book to locate the work by geographic location. The base for this type of index is a sketch map of the project to approximate scale. In this example the user, knowing the approximate geographic location of the work, is referred by the index to the proper pages for both the level runs and bench mark descriptions.

In addition, in this example the party chief has indicated, by numbering the level runs, the adjustment sequence he had in mind when planning the layout of the work.

Using this type of index does not eliminate the requirement of also keeping a simple chronological index in the front of the book. In the example you will note there are ten pages (70 to 79) left in the book; these may ultimately contain work, possibly on a different job, and will require indexing in chronological sequence.

Split book indexes

In Figure 7 the chaining was kept in the front part of the book and the angles in the back portion. This type of index separates the two operations and illustrates an exception to the chronological rule. Note that the angle at P.I. 6 (pg. 60) was measured before the angle at P.I. 3 (pg. 63). Hence the page number column is not in numerical sequence, but the index is arranged to maintain the continuity of P.I. numbers, thus making the index easy to follow.

Double indexes

In Figure 8 the work is subdivided into arbitrary units of two-mile increments, not necessarily a day's work but a convenient unit that keeps the index from becoming too long and, at the same time, does not place too many pages in each division.

The office engineer using the book is probably most interested in finding the description and elevation of a specific bench mark; the supplemental index on the right-hand page allows him to do so rapidly.

This example assumes that the bench marks have

been designated by a stamped mile number, although any numbering system could have been used.

Note that the spur line and transfer work are set apart by blank lines and can be quickly identified.

Comments. The examples shown are typical of common indexes but should not be considered the only acceptable arrangements. You may arrange the index in any of a number of ways, providing you obtain compactness, continuity, and clarity.

2-3 TITLE PAGES

General

Title pages serve two purposes. First, they divide a field book into sections for easy reference and create a definite place of beginning for each new piece of work. Second, they provide a place for recording certain general information necessary to others using the book.

When a number of different, unrelated jobs are to be placed in the same field book, each survey should be preceded by a title page recording the following information.

(*a*) Job designation (name and/or number). For example, Smith's Subdivision—Job B-1466.

(*b*) A brief generalized geographic location of the work. For example, Tentative Tract 12345, San Diego, California.

(*c*) Work description, what was done, and specifically where. For example, "Street cross sections for Bothwell Road—Station 0+00 to 21+50 incl."

If there is a possibility of confusion in the work description of the exact location, you may have to clarify with a vicinity sketch.

(*d*) References, what data you had to work from. For example,

Centerlines—C.E.F.B. 572 pgs. 26–29 incl.
Tract Map 12345.
B.M.'s—C.E.F.B. 563 pgs. 22, 28, and 30.

On large jobs a whole book or a series of books may be assigned to the job. Then the job designation and geographic location, with a general description of work contained in the book, are placed in the front of each book before the index, and the specific work description and references precede each new work process. This divides the book into kinds of work done on the same job. When, on a large job, one book is assigned for control levels only, the job designation, geographic location, description, and references should all be placed in the front of the book preceding the index. The specific work descriptions are used to subdivide the book. For example, "Levels run along west bank San Joaquin River, Kings Bridge to Modesto Wier."

The important thing is that this information is essential to the person using the book and must clearly convey the message to him.

Subdividing the book makes it easier for the user to follow the continuity of the survey. It facilitates indexing, avoids repeating and overlapping of notes, and, in general, provides a neater presentation.

In addition to this basic information which should always appear on the title page, it is sometimes convenient to place other information on the same page. If the following data will not change during the job, the title page is an appropriate place for

(*a*) Date.

(*b*) Names and duties of party members.

(*c*) Instrument name and number.

(*d*) Chain, length for a given pull and temperature, and number.

(*e*) Weather note.

If these items are not placed on the title page, it is mandatory that they appear at the top of the first page of each day's notes. The importance of these items should be self-evident; your office wants to know when the work was done, who did it, and what instruments you were using. Weather can be a very important factor in evaluating the results of your survey. The person using the book would expect different results in triangulation closures between work done on a hot, hazy day, and work done on a cool, clear, overcast day.

The title page is also an appropriate place for general notes or comments that affect the whole job. (See Figure 12, right-hand page.)

Title pages for a complete book

Figure 9 shows the title page for a complete book which must record the following.

(*a*) Job designation (name and/or number). For example, Statler Hotel, Job 555.

(*b*) Geographic location. For example, S.W. corner Figueroa Street and Wilshire Blvd., Los Angeles, California.

(*c*) General description of work contained in book. For example, construction control and layout.

Subtitle pages

Figure 10 illustrates a typical subtitle page further subdividing the book illustrated in Figure 9. It records the following.

(*a*) Specific work descriptions and specific location. For example, "Construction bench mark levels around perimeter of project setting T.B.M.'s."

(*b*) Reference material used.

The remainder of the required data, that is, date, names of party members, instrument, and weather will appear on the first page of each day's work.

Title pages, job of one-day duration

Figure 11 illustrates a title page for a job of one-day duration and records all the required data on one page.

This type of title page is applicable for small, unrelated jobs entered in a field book chronologically. Should the job run several days, this same form is applicable, providing the date, weather note, and the party members' names are entered on the first page of each day's work. You will observe that under reference 1 a complete description of the plans is noted. The importance of this cannot be overemphasized; construction drawings are subject to changes, modifications, and revisions. Therefore, a party chief, if he is to avoid being blamed for errors or changes over which he had no control, must abstract the following information from construction drawings furnished for the work and place it in the notes.

(*a*) Title of the drawings.

(*b*) Name of the person who prepared the plans. (It may be necessary to consult him later.)

(*c*) Record whether the set is signed, unsigned, or stamped preliminary. (If the drawings are not signed or are marked preliminary, it is necessary to record the name of the person who authorized you to use them.)

(*d*) The number and/or date of the last revision of the set of working plans furnished you. Errors in the plans may be discovered after you do your work and the drawings subsequently revised. If you have not noted the latest revision number, it may appear that the plans were correct but that your work was in error.

(*e*) Total number of sheets in the set. (Sheets may be added to the set after you do your work.)

(*f*) Specifically which sheets in the set contained the information necessary for you to do your work.

When there is a conflict on the plans, which frequently happens, and someone in authority at the job instructs you to hold certain data and disregard other data, place a statement in the notes setting forth the pertinent information with the name, position, and job affiliation of the person issuing you the instructions; then sign it, and ask him to sign it. It is likely that he will refuse to sign it and will tell you not to do anything until he can get the plans corrected.

On all surveying, and particularly construction, layout, the party chief has a very responsible position; he should make every effort to clarify and protect his position. An excellent rule on construction surveys is to trust no one.

Title page, jobs of longer duration

In Figure 12 the title page records the job designation, geographic location, work description, and reference material. The remainder of the information appears on the first page of each day's work. Several notes pertaining to the whole job are recorded on the right-hand page. This is a good place to record general notes or explanations, for they are immediately brought to the attention of the person using the book.

Observe reference 4; it is just as important as any other material you were given to work from and should be recorded as reference material.

Title pages, combined with sketch

Figure 13 is an example of a typical title page with a vicinity sketch to clarify the work description—specifically, which footings are involved.

References 1 and 2 in Figure 13 are typical situations frequently encountered in actual practice.

Often you will be handed one sheet of plans, not signed, with the title block incomplete, and be told to use it. When this happens, protect yourself by recording the name of the person who supplied the plan and instructed you to use it (on occasion this may be your boss; even so, record the situation).

In reference 2, the takeoff bench marks came from an unidentified field book, and conditions required you to use them; to do so is poor practice, but if this is what you did, be honest and record it.

Comments. When using loose-leaf notes, you may sometimes be supplied printed title pages with blanks to be filled in by you in the field. This is an acceptable procedure, but should the printed form not include all the data required for title pages, you are not relieved of the responsibility of recording it as would be done in any professional job of notekeeping.

2-4   PAGE HEADINGS

General

Page headings are brief statements placed at the top of a page and are intended to advise the person using the book what the page contains.

It is not necessary to place a heading on every page. For example, when running levels it is sufficient to place the heading on the first page of a certain loop or level run, thus subdividing the work for easy reference and continuity.

Page headings should be lettered large enough to be prominent and kept as brief as possible. They should not attempt to describe everything shown on the page but should be limited to the primary data.

For example, the page heading for the notes shown in Figure 25 reads, "Mon. location along Beverly Blvd." The notes actually show, in addition to the monument locations, some centerline measurements and fences, but the primary consideration is the monument location.

There are many examples of appropriate page headings illustrated throughout the sample notes.

2-5 REFERENCES AND CROSS REFERENCES

General

References and cross references are written notations referring the user to information shown elsewhere.

References are a means of sending the person using the notes to information not contained in the notes, but, in the process, the material referred to becomes a part of the notes.

For example, if you are comparing a distance measured by you with a distance shown on a subdivision tract map, the reference should read: "50.00′ record" (record refers to a recorded instrument and, in this example, a record subdivision map). Presumably the record map is completely described on the title page under references. If it is not, the reference should read: "50.00′ record, Tract 12345." In describing a found U.S.C. & G.S. monument, the note could read "Fd. USC&GS station Butte, in good condition, as described in Spec. Pub. 202, pg. 156." This reference saves writing several paragraphs.

All the reference material shown on your title page becomes a part of the survey notes because you referred to it, and the person interpreting your notes must have the same material available to make a correct interpretation.

Cross references are a means of referring the person using the notes back and forth within the notes and are used to maintain continuity.

A cross reference usually requires two entries in the notes, one at the location you are leaving and one where you are continuing.

If you are leaving page 15 and continuing on page 16, the cross reference on page 15 should read "Cont. on pg. 16," and the cross reference on page 16 should read "Cont. from pg. 15." (See Figures 22 and 23, top and bottom of left-hand page.) The fact that the work is continued on the following page does not eliminate the need for cross references.

The double entry is not required when the reference is not to the continuation of the work but is a direct reference within the notes. For example, "T.B.M. 35, see pg. 26 for elevation & description," or "See Book 10, pgs. 46–52 for traverse angles."

Comments. References properly used are great time and space savers and improve the clarity, completeness, and continuity of the notes.

The sample notes throughout the manual contain many examples of references and cross references, and it is suggested that particular attention be given to them as you study the sample notes.

PART 3: abbreviations, symbols, and codes

3-1 ABBREVIATIONS

General

An abbreviation is a shortened form of a word or phrase and is used in notekeeping to save time and conserve space.

The Appendix is a list of accepted abbreviations used in surveying notes. You should supplement this list with additional ones peculiar to your firm or area.

When a term or phrase is to be used only once or twice, it is better practice not to abbreviate it.

If a term or phrase is to be used a number of times, and you do not have an accepted abbreviation for it, show in your notes the abbreviation you have adopted. For example, O.S.W. = Ornamental Stone Wall. (See Figure 14.)

Comments. The professional acceptance of surveyors has been greatly restricted because they often use incorrect nomenclature and incorrect spelling.

In labeling cultural and planimetric objects, every effort should be made to use the correct names. If you do not know the correct name of the object and cannot find the name by asking, describe it accurately. Do not guess what it might be called.

If you are a poor speller, carry a pocket dictionary and use it. When your notes contain incorrect labeling and incorrect spelling, the person using them soon begins to wonder if other things, such as angles and distances, are also incorrect.

3-2 SYMBOLS

General

Symbols are usually thought of as graphic representations of features and objects which, when viewed, convey an impression of the feature or object being symbolized.

When using symbols special care should be taken to be sure that the symbol will convey the correct message to the person using the notes.

If the symbol you use is not commonly accepted by the company you are working for, you should be sure to identify each symbol at least once in the notes. This may be done by including one page of your notes as an index for symbols used, or by defining the symbol and object it represents in the notes the first time it is used. Take special care to select distinctive symbols which cannot be confused one with the other.

Figure 15 illustrates the use of symbols in location notes to avoid repeating the names.

Figure 16 illustrates the use of symbols to avoid repeating dimensions and naming.

There are other kinds of symbols which, through long usage, have come to have certain accepted meanings, although they are not graphically representative. For example, = is universally accepted as meaning "equal."

Figure 17 lists symbols that are generally accepted in surveying and are very useful in notekeeping.

3-3 CODES

General

A code is a system of writing in which letters, numbers, colors, and special marks are arbitrarily given certain meanings.

The keeping of field notes may, to a large extent, be considered a code system. Anyone familiar with surveying realizes that in level notes the columns marked + and − represent the rod reading for the backsights and foresights. This is an accepted way of recording information that is understood by surveyors; in other words a code. This type of coding is illustrated by many examples throughout the manual.

Special codes may be designed to solve a particular problem of briefly and neatly recording information that would otherwise become crowded and confusing to read. They may consist of using letters of the alphabet, numerals, colors, or distinguishing marks of some type.

Since this type of coding is not standardized, and as a matter of fact is not even similar in different companies, you must include in your notes a key to the code so that another person can use it.

For example, you may be locating a number of trees by plane table methods and classifying trees and noting their trunk diameters. You could use "A" for ash, "B" for birch, "O" for oak, and write the diameter preceding the letter; that is, 16"/A for 16" ash, etc., and list a key to these on the edge of the plane table sheet.

Colored pencils are often valuable in coding. You can lightly shade all rock and oil pavement, red; concrete, grey; dirt, brown; and grass, green. Colors are especially helpful in locating a number of small areas and save labeling each area, thus making the notes neat and clear. (If the notes are to be photostated, color coding should not be used.)

Figure 18 illustrates a numerical coding of the elevations along the concrete drain and retaining wall, and the notes include a graphical key. This technique has saved considerable writing and clearly conveys the position of the elevations in a compact manner.

The note below the graphical key eliminates any possible confusion whether the elevations were taken along the grid line or at right angles to the wall. The additional note concerning the elevations at the angle point in the wall further clarifies the location of these shots.

PART 4: monument descriptions

4-1 DEFINITION OF MONUMENT DESCRIPTIONS

A monument description is a written, graphical, or a combination of written and graphical, recording of the information necessary to describe the physical characteristics and the exact location of a monument and its accessories.

4-2 IMPORTANCE OF DESCRIPTIONS

It is not unusual for a law suit to be decided on the strength of the identification of a monument. Therefore it is of the utmost importance that the description clearly identify only the monument to which it refers.

4-3 ARRANGEMENT

A monument is either found or set, and the description must say which.

When describing a monument either found or set, you must answer the following three questions clearly and concisely.

(a) In what general area is the monument?
(b) Within the general area, exactly where is it?
(c) What is it?

Sometimes the answer to a is very general, relying on the answer to b to define the position clearly. For instance, on a lot survey, if you have arranged your notes as previously outlined, the title page will give a sufficient general description; for example, Lot 10, Tract 12345, Los Angeles, Calif. The survey notes, probably in sketch form, will locate the point exactly, and all that remains is to describe the type and character of the monument.

On the other hand, in writing a recovery descrip-

tion of a triangulation station, the answer to a may be quite lengthy and detailed to locate you within 25 feet of the monument, and the ties will give you the exact position.

There is no definite rule to tell you when to stop answering question a and start on b. Use your judgment to arrive at a clear and orderly arrangement.

The answer to c describes the physical characteristics of the monument—the type of material or type of object used, the dimensions, and any distinguishing facts that positively identify the point.

The balance of this part, Sections 4-4 to 4-6, divides monument descriptions into three categories which associate the descriptions with types of surveys.

It is necessary to consider all three categories as a whole concept because, as previously mentioned, there is no positive rule for the division of the descriptions and in practice you will find overlapping between the categories.

4-4 CONTROL SURVEYS

A recovery description is one that progressively localizes the position of a monument. This type of description is usually used only for monuments of a permanent nature marking triangulation stations, traverse stations, and bench marks of a higher order of work.

Descriptions of this type are written in three parts. The first two parts answer the question *where*. The third part answers the question *what*.

The first part of the description must enable a person to go with certainty to the immediate vicinity of the monument.

The second part localizes the mark by reference marks, dimensioned offsets, or detailed description. Reference marks such as these are usually set in such a manner that, should the station monument be destroyed, a new mark could be placed in exactly the same position.

The third part describes the physical characteristics of the monument with other pertinent data.

Many agencies and firms have standard forms for recording monument descriptions, and when these are furnished it simplifies the writing of the description. You can, however, prepare an acceptable description without a form if you use the following as a guide.

Before starting the description, it is necessary to record this general information for each monument.

(a) Job name and/or number.

(b) Filing or code data (sometimes a firm may have a master file system and will require special data for filing, such as approximate latitude and longitude).

(c) General designation. (Is it a triangulation, traverse, or bench mark description?)

(d) Name or number of station.

(e) Date and whether set or recovered.

(f) Name of party chief.

The description starts with a broad localization of the monument: state, county, city, and postal zone. If the monument is not in a city, the description indicates the range and township number, rancho name, or the general area by geographic name. This information is usually placed in the description heading along with items a to f, although it is actually part of the localizing description.

The remainder of the localizing description should describe clearly and concisely how to arrive at the general location of the monument starting at a well-known and well-defined land mark, such as a post office or other public building which you believe will be identifiable for many years.

Then describe the route to follow to arrive at the general location of the point, giving distance in miles and tenths between reference points and direction in compass directions, not left and right. In addition, supplemental information that will be helpful to someone retracing the description should be included, such as "Creek can be forded with four-wheel drive vehicle," or "Thence up trail 2.1 miles (a 1 hr. and 15 min. pack) to top. . . ."

The second part of the description, for a triangulation or traverse monument, gives ties to enable the exact location to be determined. Ties are also used to verify the fact that the monument has not been disturbed. For a bench mark ties are usually given in feet and tenths from fence lines, edge or center of roads, culverts, bridge abutments, etc., since it is not necessary to re-establish the horizontal position of a bench mark precisely.

The exact location data can be given in descriptive form, in tabular form, or in a sketch. The determining factor is the difficulty of describing the point precisely. A sketch is preferable whenever possible.

The third part of the description covers the physical characteristics, dimensions, and other identifying data. If you set reference marks, a full description of the type and character of each reference mark must also be included; this is also true for a subsurface mark.

This portion of the description may be placed on a sketch in the appropriate location rather than in the description.

If standard brass disks are used throughout the survey, you may note "Set std. disk in rock outcrop flush, stamped A-125." If this is the practice, a full description and a sketch of the disk must be included in the field notes.

Quite often a special book or series of books are used for all the descriptions on a particular survey, and when this is done, a detailed description of the disk may be placed in the front of each book along with any other data that is common to all points, such as "At each monument a 4"x4"x4'-0" white witness post buried 2' in the ground was set 3' north of point." This eliminates the need to repeat the information for each monument described.

When retracing and preparing a recovery description of a monument from a description previously prepared by others, you may note "Station and all ref. marks recovered in good condition as described in U.S.C. & G.S. publication 1136, description no. 7336." If there is any error in the original description, or if conditions have changed, you should make note of this—such as "The route described in 1933 adequate, but station can now be driven to with a four-wheel-drive vehicle by driving east from the 1933 'end of truck travel' for 0.3 mile to a gate, go through gate and turn north and drive along fence 0.2 mile, thence leaving fence and continuing northwest and up a ridge 0.4 mile to summit of ridge and station," or "1933 description has erroneous direction: '0.7 mile to unnamed cross road, thence east along,' should read 'thence west along'" or "Ref. mark 2 is missing, found drill hole in rock which checked tie distance and angle."

When preparing a recovery description, you should avoid lengthy writing. Accept a previously prepared description as a whole, and qualify the acceptance by noting any errors or changes.

It is often good practice when recovering brass

caps to make a carbon rub of the station mark. This is done by placing a sheet of the field book over the cap and rubbing a pencil over the page; thus you capture the image of the cap in negative and have positive proof of which monument you have recovered. Sometimes the rubbing is done on a tablet of tracing paper and cross-referenced in the field book.

When writing a description, record the necessary data as outlined. Keep in mind that the description consists of three parts and that it must enable the point to be recovered many years later. Then you will be able to adapt yourself to any form or description card and write a comprehensive description.

Triangulation descriptions

Figure 19 illustrates a triangulation description. The general information required is noted: job number, "4378"; filing data, "latitude and longitude"; general designation, "triangulation station"; name of station, "Peak"; date set, "Feb. 23, 1960"; party chief's name, "A. Jones"; broad localization, "Santa Rosa Hills Quadrangle Map, Santa Barbara County"; localization description, "Station is located . . ."; supplemental information, "Notes 1 to 3"; exact location, "ties in sketch"; physical characteristics of monument and tie points, "sketch and accompanying notes."

Bench mark description

Figures 20 to 23 represent the title page and the first three pages of a set of level notes. Some of the information required on the title page is identical with the general information required in the bench mark description and is shown in Figure 20, left. The physical characteristics and certain supplemental data are the same for all the monuments and need to be recorded only once. This information is shown on Figure 20, right. The localizing description is written in the level notes, and underlining those portions of each successive description saves time and space. For example, in Figure 23 the description for B.M. 3-20 uses the underlined portion of the previously described B.M. 3-17 and B.M. 3-19, eliminating rewriting for each description. This code is explained on the title page, Note 5. The exact ties for each bench mark are, of course, included in each description.

You will observe that the level notes begin with a two-peg test of the instrument proving it was in good adjustment before the work began. A similar test would also be included at the end of the work.

To place the following in a level book, B.M. 2″ I.P., is valueless. No one but you could find or positively identify this bench mark, and it is doubtful that you could do this six months after it was set. In the bench mark description example given, each monument can be located and positively identified by the use of title page, general notes, coding, and additive descriptions to describe jointly a series of bench marks. The information recorded satisfies all the conditions set forth in this section, and you can easily transcribe this to any type of description card or form.

4-5 ENGINEERING SURVEYS

For the purposes of this discussion, engineering survey monuments refer to all monuments other than those marking boundaries or permanent marks for points of control surveys. It is assumed that the data shown on the title page localizes the point to the extent that the first part of the description does not have to be repeated for each monument, and the person using the notes will know the general area of the job. The problem here is to localize it within the job area and then describe the point for positive identification.

For example, when running construction levels over an area before starting construction layout surveys, the descriptions in the level book must be sufficiently clear to allow someone to recover the monument using only the title page and the description of the point. On a construction job care must be taken that the points referred to in the description will not be destroyed during construction.

This type of description is similar to the previously discussed control monument descriptions, but it need not be as detailed, as long as the point can be recovered and positively identified. For example, if control traverses have been staked, you can refer to them in subsequent descriptions: "B.M. 21, set 2″ I.P. flush, 12.5′ right (SW) of station 12+65.21, Road 'A' traverse, Job 15732, Book 3, pg. 21." This description ties the bench mark to the traverse, and it can be found by referring to Book 3, page 21, locating station 12+65.21 and searching 12.5′ right.

When establishing points on construction, care must be taken to insure positive identification of the point referred to. For example, if you set a spike in a power pole, be sure you identify the pole by number, state the height of the spike above ground, and state on which side of the pole the spike was set. Preferably use a type of spike that is not commonly found around construction. It is possible that another survey party has set a bench mark on the opposite side of the pole, and if your description is not clear, someone can take off the wrong spike. It is good practice to circle the point with paint and to paint the identifying number on the pole for positive identification. Do not attempt to paint the elevation on the pole until you have closed and adjusted the level run and are positive of the elevation. If you do identify with paint, say so in your notes.

When starting your work at found points on construction surveys, identify the point in your notes and give the book, plan, or map reference you had for the authority of the point.

In setting points for construction, it is very important that the point be adequately tied out. Points are easily destroyed during construction, and it is especially important that the elevation of the point be properly referenced. To note "Set 2"x2" with cup tack 16" deep," has less meaning on construction than on other types of work because grades are probably being changed and the reference "16" deep" could easily be "46" deep" a day or two later. It is better to note: "16" below finish floor elevation of Warehouse B." You can obtain this reference with a hand level, and it is a positive vertical reference which is not subject to change.

It is important that you put in your notes how you marked the identifying laths, especially when marking cuts or fills on laths. Contractors move laths from point to point indiscriminately, and you should never mark a cut or fill on a lath without also marking the horizontal position on the lath that the cut or fill is intended for, that is, the station number and left or right offset. Be sure this is reflected in your notes; it may save you embarrassment later.

Construction notes

Figure 24 illustrates notes for a foundation layout survey. In this example, all general locations of monuments and localizing descriptions would be placed on the title page, leaving only the exact location and the physical description of monuments to be recorded on the layout sketch.

The angles and distances shown on the sketch give the exact location; and Notes 1 and 2 give the physical description of the monuments found or set and the information marked on the laths.

This figure is also an example of a common situation wherein the lot corners as actually found do not agree with the plans furnished and the party chief has been instructed to alter the layout from the position indicated by the plans. This is explained in Note 4, and the discrepancy is indicated on the sketch by recording the angles and distances shown on the plans, and the measured angles and distances.

Observe the use of coding for compactness, and note the diagonal check distances to prove the building layout is 90°. Also note that there was not enough space to complete Note 4. It is properly cross-referenced to the left side of the page rather than crowding the lettering.

4-6 BOUNDARY SURVEYS

Boundary survey monuments refer to all monuments used in connection with cadastral surveys.

It is again assumed that the title page localizes the point so that the first part of the description does not have to be repeated for each monument, and the survey field notes give an exact location for the point. The physical description is substantially the same as it is for control and engineering monuments, with added emphasis on evidence observations that may assist in substantiating the record authority and legal acceptance of the point.

When describing a monument either set or found on boundary surveys, the description should always include the following.

(*a*) Did you set it or find it? If you found it, is there authority for the point such as a City Engineer's field book or a tract map?

(*b*) Size of monument. This is generally restricted to the nominal cross-section dimensions, such as 2"x2" stake or 1"x2" stake; for pipes you should specify whether it is inside or outside diameter, such as 2" I.D. pipe or 2" O.D. pipe; for stones usually three dimensions are given, 12"x16"x30" stone, and the cross-section dimensions are the upper end on the stone (12"x16").

(*c*) Type of monument. Is it a pipe, stake, stone, boat spike, etc.?

(*d*) Does the monument have a precise point, such as cut cross, cup tack, wood plug with no nail, open pipe, drill hole, etc.?

(*e*) Relation to natural ground, such as flush, 12" deep or 12" subsurface, 3" above ground, etc.

(*f*) Condition of a found point. Is it rusted, rotted, badly rusted, leaning, loose, disturbed, in good condition, etc.?

(*g*) Supplementary information, if any, such as brass tag LS 2333, washer, shiner, 4"x4" witness post 2' north, 36" lath tied to pipe marked NW Cor Lot

2, red flagging nailed to fence, fence keel marked 10' offset, in a mound of rocks, relation to adjacent fences, etc.

In addition, if you set or find ties you should make note of this. For example, "Set 4 nails straddle points-tangent-2' offset," or "Set 4 nail swing ties-2' offset," or "Checked by C.E. ties F.B. 22222-77," or "Found 30" blazed oak tree, 30.7' NW" (if you have ties of course you should always use them as a check).

The following descriptions are written in good form, and you will note that each description contains the elements mentioned.

1. Found as per Tract Map 12345, 2" O.D. iron pipe with redwood plug, tack and disk L.S. 2333, 6" above ground and leaning NEly, appears to have been disturbed.

2. Set 2"x2" redwood stake flush, with nail and disk L.S. 2333, and 24" lath and flagging marked "NW Cor Lot 17."

3. Found 6" O.D. well casing, concrete filled, with brass cap stamped "Surveyed by A. B. Jones—1895." Top, 14" above ground. Also found remains of several sight poles and guy wire, point apparently accepted by other surveyors as corner. Mon. is in good condition.

4. Set 12" bridge spike flush in pavement, with disk L.S. 2333 and shiner, punched spk. for exact point. Also, set two nails for north-south straddle point 2.50' tangent offsets. (See pg. 20 this book for sketch of additional ties and offset to three found monuments.)

5. Set temporary point (nail) from ties CEFB 22222 P-7. Ties give a 0.03' diamond; used center for true point. Conc. is broken and C.E.'s L. & T. is lost.

6. Found chiseled cross in 10"x15"x30" granite stone set in center of 3'-dia. rock mound. Top of stone 24" above natural ground. Mon. seemed undisturbed. No evidence of pits called for in G.L.O. notes.

7. Found old rotten 2"x2" redwood stake 3½' deep with evidence of rust. Accepted center of stake and set 24"x2" O.D. pipe with plug, nail and disk LS 2333 over same and flush with natural ground. I did not disturb the found stake.

These descriptions may seem lengthy, but to leave out any part of them would be to keep incomplete notes and might well lead to your being asked such questions as "What did you find here?" "What was the condition of the stake?"

Figure 25 illustrates the recording of the location and description of existing monuments. Note that by using a numerical code considerable space is conserved. In addition, descriptions 1 and 3 each apply to two found monuments of identical type, and it is not necessary to repeat the descriptions.

The enlarged detail clarifies the location of several monuments in close proximity to each other.

The authoritative reference is given where applicable.

The assumption in this example is that the title page gave the general information and the localizing data, and therefore only the exact location and the physical description of the monuments need be shown on this page.

Note that certain other data pertinent to the establishing of boundary corners is also shown, for instance, station numbers of fence and building walls.

Comments. As previously stated, no positive rule exists for the division of descriptions, and in practice you will have to adjust the descriptions to satisfy particular conditions. For example, you may be doing triangulation, establishing control for mapping and also for a boundary survey. If you elect to use a 2" iron pipe found at what appears to be a ranch corner for one of your triangulation stations, the first and second parts of the description will follow the form for control monument recovery descriptions, and the third part will conform to the requirements for boundary survey monument descriptions. That is to say, the third part will not only describe the exact physical characteristics of the pipe, but will also note such facts as the existence of old 1"x2" straddle points, the exact relation of adjacent fences, the location and description of any other survey monuments found near by, a reference to record authority if appropriate, and a verification with bearing trees, etc.

The fact that you found a monument in a suitable location for a triangulation station and suspect it may be a property corner does not necessarily mean that it will finally be accepted as the ranch corner. It will have to be proved by additional boundary surveying. You should, however, observe and record any data that may shed light on the reliability of the point at the time you write the description.

PART 5: recording support data

5-1 GENERAL COMMENTS

Support data is information which is not normally associated directly with the actual accomplishment of a survey but which will establish the user's confidence level in the survey.

The type of information that constitutes support data is quite varied, and the selection and advisability of recording it are, again, a matter of judgment. It may appear in the form of a brief informative notation, a lengthy narrative of an event or situation, an oath, a photograph, standardization notes, or some other form.

The criterion should be whether the information is necessary to establish an acceptable confidence level. For example, it certainly is not necessary to support a low-order location survey with standardization data on the tape. On the other hand, on a triangulation base line, to exclude this data would reduce to zero the user's confidence in the work.

The remainder of this section covers examples illustrating the recording of support data.

5-2 PROCEDURES, OBSERVATIONS, CONCLUSIONS, AND SITUATIONS

These are included in the notes in narrative form, kept as brief as possible but still conveying the message clearly. They should be headed with an appropriate title.

In Figure 12 the special note at the bottom of the right-hand page is an example of recording a situation that could well affect the survey.

Figure 26 illustrates the recording of a procedure, observation, and conclusion, omission of which would make the interpretation of the survey impossible.

5-3 OATHS

A number of states grant a surveyor the right by law to administer and certify oaths. For example, in California under Chapter 15, Division 3 of the Business and Professions Code, known as the Land Surveyor's Act, Article 5, Section 8760, a licensed land surveyor may administer and certify oaths under certain conditions. The following is a copy of Section 8760.

Every licensed land surveyor or registered civil engineer may administer and certify oaths:

(a) When it becomes necessary to take testimony for the identification or establishment of old, lost or obliterated corners.

(b) When a corner or monument is found in a perishable condition, and it appears desirable that evidence concerning it be perpetuated.

(c) When the importance of the survey makes it desirable to administer an oath to his assistants for the faithful performance of their duty.

A record of oaths shall be preserved as part of the field notes of the survey and a memorandum of them shall be made on the record of survey filed under this article.

The following definitions are pertinent to the understanding of recording statements and administering oaths.

1. An oath is a ritualistic declaration, based on an appeal to God, that one will speak or has spoken the truth.

2. A deposition is a written statement by a witness made under oath.

3. To depose is to state something under oath.

4. An affiant is a person who makes and affidavit.

5. An affidavit is a written statement made under oath.

6. The symbol ss means "namely" or "as follows."

A licensed surveyor may have occasion to take a deposition of a person or persons regarding any of the three situations *a*, *b*, and *c*. Before the individual signs the deposition, the surveyor must administer the oath by asking the individual the following question:

"Do you solemnly swear (or affirm) that the statement you are about to sign is the truth, the whole truth, and nothing but the truth, so help you God?"

Or, less formally, you can ask him

"Do you swear this is a true statement?"

After an affirmative answer is received, the affiant signs the deposition and the surveyor certifies to it.

The accompanying form is one that can be used for recording testimony of this type:

STATEMENT PURSUANT TO AUTHORITY UNDER CHAPTER 15, DIVISION 3, ART. 5, SECTION 8760 OF THE CALIFORNIA BUSINESS AND PROFESSIONS CODE

State of _____ } ss
County of _____

I, _____, being first duly sworn, do depose and state the following:

_____
Signature of Affiant

Subscribed and sworn to before me, a licensed surveyor in and for said State, this _____ day of _____, 19____.

_____
Signature of Licensed Surveyor

_____
L. S. Number

Figures 27, 28, and 29 illustrate typical situations that might require the administration of an oath.

Considerations to bear in mind when preparing depositions are

(*a*) Stick to facts. The statement the affiant signs should contain only statements that he is sure from his own knowledge are true.

(*b*) The oath must be administered; otherwise the affidavit can be challenged as not being made under oath.

(*c*) Any party chief can prepare such an affidavit and have it signed, but unless he is a licensed land surveyor or registered civil engineer, he probably will not be able to have the statement admitted as evidence in court. If you are not licensed, you can prepare the statement and have the affiant subscribe and swear to it before a notary public.

5-4 ADJUSTMENT, STANDARDIZATION, AND CALIBRATION

Adjustment notes

A party chief is expected to be able to adjust the adjustable parts of surveying instruments. These are the parts that are altered by normal handling of the instrument and temperature changes.

If you expect to do accurate work, you must test your instruments at frequent intervals and make whatever adjustments are necessary. It is particularly important to check the instruments at the beginning and end of a survey if a slight unadjustment would affect the desired accuracy of the survey, and, of course, whenever you suspect the instruments may have been bumped or vibrated.

When testing and adjusting an instrument, it is necessary to keep notes of the results of your adjustment to prove the instrument was correctly adjusted, establishing a confidence level. A good system is to keep notes on the test, that is before adjustment, and to record the data after adjustment to prove that the adjustment was correct.

It is not the purpose of this portion of the handbook to tell you how to make the necessary adjustments; you can refer to any number of surveying texts for this information, but the following sample notes illustrate how to record the necessary data for some of the more common adjustments.

Figure 30 illustrates notes for testing and adjusting a dumpy level. The three tests and adjustments to be carried out are

(*a*) Axis of level tube must be perpendicular to the vertical axis of instrument.

(*b*) Horizontal cross hairs must lie in a plane perpendicular to the vertical axis of instrument.

(*c*) Line of sight must be parallel with the axis of the level tube (two-peg test).

In this example the tests and adjustments are made and recorded in this order. It is necessary to record for each one the test, its corresponding error, what you did to correct the error, and a second test to prove that the adjustment removed all the error.

When the data is clearly recorded, anyone looking at the notes can tell how much the instrument was out of adjustment and will be assured that the instrument was correctly adjusted.

Figure 31 illustrates notes for testing and adjusting a wye level. The following three tests and adjustments must be made, and a two-peg test must be used as a check.

(*a*) Axis of sight must coincide with axis of collars.

(b) Axis of level bubble must be parallel to axis of sight.

(c) Axis of level bubble must be perpendicular to vertical axis of instrument.

(d) Level must check by two-peg method.

The test and adjustments are made in this order, recording the test, error, method of correction, and proof.

Figure 32 illustrates notes for testing and adjusting a transit. The following seven tests and adjustments must be made.

(a) Plate-level bubbles.

(b) Preliminary adjustment of vertical cross hair.

(c) Final adjustment of vertical cross hair.

(d) Horizontal cross hair.

(e) Standards.

(f) Telescope-level bubble.

(g) Vertical circle vernier.

Note that only the first and the final tests and errors are recorded. It is not necessary to record any intermediate trials that are made to achieve correct adjustment.

Standardization notes

Standardization is the comparison of an instrument or device with some accepted or adopted standard to determine the value of the instrument or device in terms of the accepted or adopted unit.

The following information must be recorded for standardization notes.

(a) The pertinent data identifying both the instrument or device being standardized and the accepted or adopted standard.

(b) The conditions under which the comparison was made, the date, the names of party members, and pertinent data concerning any supplementary instruments used in making the comparison.

(c) The computations and the corrections, in units of the adopted standard, to be applied to the instrument or device being compared to standardize it.

(d) When appropriate, a description of the method used to arrive at the comparison.

Figure 33 illustrates notes for standardization of a steel tape. In this example the tape being standardized is described, and the base line, which is accepted as the standard, is also described with a verification note that it was "checked today with Lovar tape."

The conditions under which the comparisons were made are noted: weather, temperature, pull, and support. The names of party members and the supplementary instruments used are also noted: ther-

mometers 8 and 12, spring balance 4, with a note that all were known to be correct.

In this example two comparisons were made, one with the tape fully supported and one with the tape supported at each end. Four readings were taken for each comparison, and the pertinent data and computations were recorded, resulting in the standardized length of the tape.

The method used is noted, that is, compared with a base line of known length. It is further noted that the base line length was verified on the date of the work by checking with a Lovar tape.

Figure 34 shows a set of notes for standardizing a spring balance. All the pertinent information is recorded, and the mean correction to be applied is computed.

In this example the adopted standard is the Ajax dial face spring balance, which was previously tested and found to be correct within $\pm\frac{1}{16}$ pound. The procedure is explained in some detail because there are several ways to compare a spring balance—certified weights, for example.

Calibration notes

Calibration is the determination of the values of the supplementary marks on a measuring device by mechanical interpolation based on values obtained by standardization.

When the total length of a tape has been determined by standardization, the values of its intermediate marks may be determined by calibration, assuming that the divisions on the tape are directly proportionate to the total length.

Calibration by its definition is a mathematical operation and is usually accompanied by standardization notes.

In the previous example, Figure 33, the divisions on the tape would be calibrated by direct proportion. For example, an observed 50 feet on the tape is calibrated to the standard by dividing 50 by 100 and multiplying by 99.990, which equals 49.995. Thus the observed 50 feet is calibrated to the standardized length of the tape.

If the supplemental divisions on the device are not directly proportionate to the total length, or if the mechanical operation of the device is not directly proportionate, calibration is not applicable and it is necessary to standardize the intermediate marks. In the example of the spring balance, Figure 34, the intermediate marks are probably proportionate, but the expansion of the spring is not. Therefore each 5-pound increment was standardized, and between 5-pound increments it may be assumed that the readings are proportionate, at least to within acceptable accuracy tolerances.

Figure 35 illustrates notes for determination of the stadia factor. In this example the stadia hairs, rod graduations, and effects of observing are all standardized as one operation at ten locations over a 1000-foot range. The adopted standard is the measured distance, and at each of the ten locations a stadia factor is determined.

The mathematical mean of the individual stadia factors is in effect a calibration of the whole procedure. The assumption is that any intermediate positions will be in direct proportion to the mean stadia factor.

The field notes in this example are arranged in an orderly manner. All the necessary data for standardization is recorded. The description of the method used is brief, for it should be self-evident that a line was measured and stadia readings were taken over the chaining points. The mathematical calibration formula is shown and sufficient data tabulated to allow the work to be checked.

Figure 36 shows notes for the calibration of an echo sounder used in hydrographic surveys. In this example the adopted standard is hand sounding the depth with a lead line, and the comparison is made after applying the constant C, at random depths between 12' and 60' of water. The correction K is computed for each comparison and the mathematical mean computed. The notes are arranged to record all the data for each comparison and to allow a ready review of the work.

In hydrographic work there are many sources of small errors of about ± 0.2 foot that affect the depth determination, such as roll of the boat, slope of the bottom, accuracy of the chart reading, etc. Therefore the standardization comparison will be plus or minus some small amount.

The calibration determination is made by inspecting the K's, and if the variation from the mean is a reasonable amount, it is a valid assumption that the intermediate values will be directly proportionate.

You will note that all the pertinent data is recorded and that tests were made both before starting the work and after completing the work to prove that the instrument stayed in adjustment.

In hydrographic work it is customary to record the length, description, and name of the vessel used, as well as the location of the transducer head.

Comments. Many firms have test ranges established near their offices and test and adjust the instruments as a regular procedure. Some maintain special field books or forms for recording the adjustments and standardizations, and this procedure should not be duplicated unnecessarily. If you have occasion to adjust, standardize, or calibrate your instruments, however, the operation should be recorded in the field notes.

5-5 PHOTOGRAPHY

Properly used, photography is an excellent method of recording many types of survey data and is exceptionally suitable for support data. Many survey parties carry a Polaroid camera as standard equipment and use it to record supplemental and support data.

When a photograph is taken as support data, the position and direction of the camera must be shown in the notes. Each photograph should be numbered and cross-referenced in the field notes. A list and brief description of all the photographs taken as part of the survey should also be included in the notes in case the photographs later become separated.

In addition to numbering the photograph, indicate on the back of it the job number, date, the name of the person taking it, and a cross reference to the field book and page.

Figure 37 illustrates field notes supported by photography showing additional data. In this example the validity of the establishment of the ranch line is important to the survey. The photographs support the written description of the monuments and bearing tree. In such a survey the confidence level might well be a major factor in influencing a court's decision.

PART 6: typical arrangements of notes

6-1 GENERAL

As previously stated, it is not the intention of this handbook to develop a set of standard forms for note-keeping. Therefore the examples given in this section should not be considered rigid. They are intended to show enough examples of notes for a variety of survey situations that, taken as a whole, illustrate how notes may be prepared to fit any survey conditions encountered.

6-2 TAPING NOTES

There are a number of ways of recording taping notes depending on the type and magnitude of the survey.

It is usually better to keep measured distances and horizontal angles on separate pages, for this keeps the work from becoming crowded and results in a much clearer presentation. It is perfectly acceptable, however, to keep them together if it can be done neatly and clearly.

Short distances (graphical form)

The usual practice when recording data involving a number of short measurements is to use a sketch and graphically position the measurements.

Figure 38 illustrates notes recording a number of offsets to locate found monuments. Observe that the offsets are taken at right angles to the center lines, and the sketch is of sufficient scale to show graphically that offsets are at right angles without having to add 90° symbols. The descriptions of the monuments are kept away from the body of the sketch for clarity, the authority for each point is given when known, the fact that the city engineer's ties did not result in a common point is mentioned

(Note 1), and there is an explanatory observation about the pavement condition (Note 2).

Figure 39 shows a center line tie sheet. The tie points are referenced to the BC's of the curbs and to the center line produced, to assist anyone using the work to locate the tie points quickly.

The sewer manhole is detailed for clarity, and the throw-over symbol indicates that the points are on a straight line; the indication of the top of curb needs only be shown once and is not repeated for each tie point. Note 2 states that the measurements are standard at 68°, and the page is properly cross-referenced for the authority of the center line intersection monument (Note 1).

The tie angles are shown, since it would not be practical to prepare a separate sketch showing only three angles.

It is not necessary to arrow the throw-over ties because it is perfectly clear what points they refer to.

Figure 40 illustrates notes for a foundation location involving a number of short measurements. In this example the boundary was surveyed previously but sufficient measurements are shown to check the original monuments for movement.

The sketch is large enough to show clearly that the offsets were at right angles to property lines.

The location dimensions are totaled on the left-hand side of the page as a check on the work.

The note inside the foundation clearly states where the dimensions were taken, and Note 1 clarifies the fact that the offsets were 0.2′ from building corner. (This is done to avoid the irregularities usually occurring at the corners of concrete.) No angles were turned, but the building was checked for square by measuring diagonals (Note 2).

The offsets are shown as actually measured. The building front is shown 2.10 feet from the lead and tack offset line instead of 0.10 feet from the property line.

Long distances (graphical form)

The main problem when distances are large is that of showing the chaining with the proper corrections and reductions without having cluttered and confusing notes.

Figure 41 illustrates notes involving a number of long distances. The chaining notes showing measurements, temperature, pull, and reduction are recorded on the left-hand page, and the adjusted distances are shown in the appropriate location on the sketch. The two are numerically cross-referenced, a process which also indicates the sequence in which the measurements were taken.

All the measurements were taken with a 100′ tape, breaking chain only for the last part of each distance; by this method the whole distance can be recorded on a single line. If it were necessary to break chain several times on a line, several lines would obviously be required for each total distance.

The left-hand page has subheadings to separate the work by streets for clarity. The authority for each point found is noted, and the method used to establish each point set is recorded.

Observe the note on Block 21 "(Cont. on pg. 50)." Page 50 would show an enlarged sketch of Block 21 of sufficient size to continue recording the survey data without becoming cluttered.

It is not necessary to show the calculations to reduce the chaining to standard at 68° since it was probably done on a slide rule, but you must record the temperature and pull so your answer can be checked.

Figure 42 shows notes for measuring long distances where the total distance is composed of several increments, some involving slope chaining.

In this form the sketch is kept on the right-hand page, showing only the total adjusted horizontal distance between P.I.'s and leaving ample space for other information. The detail of each measurement is recorded on the left-hand page.

Note that column headings are selected and arranged to define clearly each measurement recorded with the supplemental data necessary to reduce the distance to a standard measurement. Observe the cross reference to the page showing slope reduction calculations.

The correction for temperature·was calculated on a slide rule and is shown in the right-hand portion of the temperature column.

Long distances (described form)

The descriptive form of chaining notes is applicable for running long traverses when the chaining and angles are done as separate operations.

Figure 43 shows the descriptive form of notes for recording chaining. One page is used for each tangent, and the column headings are arranged to allow space for recording all the necessary information with the adjusted distances. The distance is totaled at the bottom of each page.

The intermediate chaining points are numbered for future use.

Observe Notes 1 to 3 stating what was done, what was set, and how the intermediate points were marked. Note 2 states clearly that the chaining points were not exactly on line (honest notes).

Stationing (graphical form)

In this form of notekeeping, the chaining must be done in a manner that allows the tape to accumulate the stationing. Then it is only necessary to record the station numbers.

Generally, in this type of a survey, no attempt is made to correct for temperature or standardization, but sometimes these are approximated by an addition or cut at the zero end of the tape. If this is done it should be so noted.

Figure 44 illustrates graphical stationing notes for a traverse and location survey. In this form the sketch is arranged to allow the stationing to increase from the bottom to the top of the page; thus right-hand offsets in the field are plotted on the right-hand side of the traverse line.

Note that the station numbers are placed parallel to the offset dimension lines and not at right angles to the traverse. This insures an orderly presentation.

In this form, as in all location notes, sufficient data must be recorded to allow a positive plotting of the objects being located. For example, at Station 16+10.7, without the notation "(Barn face, prod)" it would be impossible to plot the barn with certainty.

The P.I. angles can also be shown if they will not clutter or be confusing.

Stationing (descriptive form)

Figure 45 illustrates notes for laying out a pipeline from plans.

In this form, as in Figure 42, the chaining must be done by a method that will accumulate stationing with the tape.

The column headings are arranged to show the station number and allow for recording offsets and deflection angles, and on the right-hand page the appropriate notes are made: type of monument set, how the laths are marked, and the location data.

Observe at Station 18+67.30 that both the actual deflection angle to join the existing pipe and the angle shown on the plans are recorded. In addition, at the join, Station 20+11.76, the actual measure station number is shown and the plan station number is noted. This is an example of keeping honest notes. Record actual measurements, not what they are supposed to be.

The same form can be used with station numbers increasing from the bottom to the top of the page; either way is acceptable.

Figure 46 illustrates stationing notes of a higher order of chaining. The form is arranged to record each distance measured with the pertinent information, and the adjusted distances are accumulated to arrive at the station numbers.

Note that the first column on the right-hand page is for recording stadia distances to be used as a check on the chaining.

The horizontal angles for this traverse would be recorded on separate pages, because to include them on this form would crowd the work, especially if several sets were to be turned at each P.I.

Observe that the description of each point set also includes the length of the lath and the way in which the lath is marked.

6-3　ANGLE NOTES

There are a number of ways to record angle notes depending on the type of survey and the number of sets of angles being turned.

As previously mentioned, it is usually better to keep measured distances and horizontal angles on separate pages.

Random angles (graphical form)

Random angles are usually recorded in graphical form because it is simpler to do this than to describe the situation for a few angles.

Figure 47 shows notes recording the angles taken to locate pilings by two-point triangulation intersection. The angles are turned, once direct and once reversed, and the mean is computed. A numerical cross reference is used for clarity, and only the mean angle is recorded in the appropriate position on the sketch.

Traverse angles (graphical form)

In Figure 48 the traverse has been previously chained and monumented by others, and the angles are recorded in graphical note form.

It is not necessary to use a numerical cross reference for the angle sets because the arrangement makes clear which set of angles refers to which point in the sketch.

In this case the angles were repeated four times, and the accumulated values for the first, second, and fourth were recorded. The mean was computed and placed in the proper location in the sketch.

Note that the descriptions of the P.I. monuments are recorded in full, even though they have been previously set and described in a separate book as part of this survey. This is done for positive identification of the point occupied.

Traverse angles (descriptive)

Figure 49 illustrates traverse angles turned six times with a repeating instrument. The accumulated values for the first, second, and sixth were recorded.

This form uses one page for each point occupied, with the P.I. number and description of the monument placed at the top of each page. The rest of the page is arranged to record the angle first, the horizon closure next, and other pertinent notes at the bottom.

Observe in this example that when the surveyor was at P.I. 36 he noticed that the signal pole at P.I. 37 appeared to be leaning and recorded this; at P.I. 37 the amount and direction of the leaning signal was determined and noted in sketch form at the bottom of the page. From this data the office can make the eccentric signal computation. The angle to the eccentric was measured twice, not to obtain more accuracy since the distance is only 0.31′, but as a proof of the correctness.

The note "guy wires taut, signal steady" is extremely important because without it the user might wonder whether the signal was loose and had perhaps shifted between the time it was pointed and the time the eccentric was determined. If the signal were loose it would have to be reset and P.I. 36 reoccupied.

Triangulation angles (graphical form)

Figure 50 shows angles for a single quadrilateral of triangulation. The description of each point is placed adjacent to the point, and the angles are placed in the appropriate position on the sketch with the means calculated.

The sums of the triangles are tabulated outside the sketch, and the errors of closure are noted for each triangle.

Complete triangulation nets can be recorded using this form, provided each figure is placed on a separate page and proper cross referencing and indexing are used.

Triangulation angles (descriptive form)

Figures 51, 52, and 53 show notes for one set of triangulation angles taken at four positions of the plate and two sets of vertical angles, observed with a T-2 instrument.

The column headings are arranged to allow recording the plate position, the object observed, the observations with instrument direct and reversed, the mean of the direct and reverse, and the reduced directions. On the right-hand page opposite the first position, the reduced directions are extended to form an abstract of directions with the mean for the complete set.

After all positions are completed and abstracted, the reobservation for any rejected directions is recorded and the new value entered in the abstract with the rejected value being crossed out.

The vertical angles are tabulated separately from the horizontal angles with the signal heights and the height of instrument noted. It is sometimes more convenient to place the vertical angles on a separate page.

A separate section of the book should be used to check triangle closures, with a diagram to show results.

In this example only four positions were taken, but the note form is suitable for any number of positions.

You will note that on Figure 51, the abstract for position 2, pointing S.R. 89, the seconds are recorded $\overline{56}$. This indicates that the minute value for this direction is one minute less than recorded in the abstract. In other words, this direction is 134°06′56″, not 134°07′56″.

The rejection limit for the directions on this job was ±04″ from the mean. In pointing S.R. 90, position 2, the direction was more than 04″ from the mean. Reobservation found the direction to be even further from the mean, indicating the presence of refraction; therefore both values for position 2 were used, and a new mean was calculated.

6-4 LEVEL NOTES

Level notes are arranged to allow the backsights and foresights to be accumulated to determine the elevations of points required, with provisions for recording clarifying notes and bench mark descriptions.

Variations in the arrangement of different types of level notes are made primarily for simplicity of recording.

Single wire

Single wire levels (reading only the center cross hair) are generally used on third-order leveling.

Figure 20, 21, 22, and 23 show a set of differential level notes, reading a single wire with alternate rods at turning points, for a third-order circuit to establish elevations on certain permanent bench marks.

The column headings are arranged to record the sight, backsight, H.I., foresight, elevation, adjusted elevation, and the distance from the instrument to the sight, with ample space remaining for the bench mark description.

The title page and method of describing the bench marks have already been discussed (see Section 4-4 Control Surveys, Bench mark description). Note that the title page is complete, supplying considerable information common to all the bench marks.

The B.S. and F.S. are accumulated to determine the elevations of each turning point and in addition are summed by columns, with the total difference in elevation being checked against the elevation difference at the end of each page. The stadia distances to the sights are recorded and totaled to show that they are in balance. The picture point descriptions are cross-referenced to the appropriate photograph.

Adjusted elevations are not entered in the notes but are left for the office to complete after making the final adjustment.

These notes conclude with a two-peg test at the end of the run to prove that the instrument remained in adjustment.

When the total level run extends over large differences in elevation, that is, several hundred feet, the notes should contain a standardization check of the rods used.

Note in Figure 23 that the signs of the two rod readings with the rod inverted are circled to call attention to the fact that they have been reversed.

Figure 54 illustrates a form for recording notes

for double-rodded circuit. The column headings are similar to those of Figures 21, 22, and 23, with the left and right pages used to record the A-rod circuit and the B-rod circuit respectively.

The bench mark descriptions are entered on only one page and need not be repeated since both circuits use the same bench marks.

It is not necessary to sum the rod reading to check the elevation differences because the H.I.'s for each circuit afford a ready check.

The field-adjusted elevation of the bench marks established is entered as the mean of the two determinations and is used to advance the level circuit. The final adjustment is made as an office procedure.

Figure 55 shows notes for recording levels using a double-faced rod. This arrangement is almost identical to that of Figure 54. The pages are headed "feet and yards." The yard readings are used as a check against misreading of the foot rod. Normally, in this type of leveling, the order of work does not require an adjustment and frequently the runs are stub-ended.

Figure 56 is an example of recording level notes where the elevations of intermediate turning points are not required and forward and backward runs are made.

The pages are arranged to sum the backsights and foresights between T.B.M.'s, the differences in elevation between the two runs are compared to determine the divergence, and the adjusted difference is used to establish the field-adjusted elevation. The final adjustment is made as an office procedure.

In this example all the pertinent information and descriptions of T.B.M.'s would appear elsewhere in the notes.

Observe the note concerning the balancing of sights.

Three wire

Three wire levels (reading all three cross hairs) are generally used for first- and second-order leveling and are frequently referred to as precise levels.

Figures 57 and 58 show notes for precise levels. The column headings for the left-hand page are arranged to show the station number (setup number), rod number, three wire readings for the backsight, the mean of the three wire readings, the interval between top and center wires and between the center and bottom wires, the sum of the intervals, and a foot-reading face on the back or edge of the meter rods to be used as a check if desired. (In this example the foot check was not used.) The right-hand page is similarly arranged, with the addition of a column for the rod temperature.

Backsights are always recorded on the left-hand page and foresights on the right-hand page.

The page heading states whether it is a forward or backward run, and the bench mark numbers are noted.

The total elevation differences and the sum of the intervals are calculated at the end of each run, with the divergence calculated at the end of each circuit.

No provision is made for recording elevations, since the field notes are abstracted as an office procedure before making the adjustment calculations.

Special level notes

This section includes sample level notes for use with special instruments.

Figure 59 shows level notes arranged to record the necessary data when a forked-hair instrument with optical micrometer is used. The column headings are arranged to list the station (setup number), the right-hand side of the rod in meters, decimeters, and centimeters, the instrument micrometer reading in decimals of centimeters, the left-hand side of the rod as a check (there is a constant offset between the two sides of the rods used with this type of instrument), the top and bottom stadia hairs, and the difference. The pages are headed backsight and foresight.

The information recorded at the top of the page on the line noted "from pg. 17" is the total carried forward from the previous page.

The columns "right" and "mic." are totaled, and the difference in elevation is computed. The total stadia distance is also summed to check the balance between backsights and foresights.

As in the previous example (Figure 58), no provision need be made for recording actual elevations since the field notes are abstracted as an office procedure.

Figure 60 shows sample notes for establishing elevations using a Lenker rod. The note form is arranged to record the actual tape readings set and read at each point required, with a column for recording the full elevation value.

The tape rod should never be used on work requiring a high accuracy because it has no target and is subject to slight errors in manipulation, but it is well suited to short runs of a few setups.

Figure 61 illustrates a note arrangement for use with Beaman stadia arc. This form is designed so that all the data concerning each backsight and foresight can be recorded on a single horizontal line.

The columns on the left-hand page are arranged to receive the data in the order in which it becomes available. For example, you know immediately if the shot is a backsight or foresight. The instrument man reads the stadia distance first, sets the arc next; the recorder calculates the product while the instrument man is reading the rod, then determines the difference in elevation and applies it to either the H.I. or the elevation of the turning point.

The entries on the right-hand page are aligned horizontally to be opposite the turning point symbol and foresight.

The adjustment, as would be expected from stadia levels, is rather large and is noted in the elevation column for convenience in arriving at the adjusted elevation.

Note that when appropriate the sign of the values in the product column is circled, indicating that they are the reverse of normal. For example, on the first line the arc reading of 38 indicates a minus product, but because the sight is a backsight the sign is reversed. This not only helps to prevent mistakes in the field calculations but also clarifies this irregularity for anyone checking the notes.

Figure 62 illustrates notes for recording barometric altimeter levels.

The left-hand page is arranged to record the station or point being measured, the time, the observed instrument reading, the temperature, the observed difference in elevation, and the mean temperature of the two successive readings. The right-hand page is arranged to make the necessary calculations and adjustments.

The calculations are done as a field procedure so that the answers obtained on forward and backward runs can be compared for reliability. (In order to obtain reliable results using an altimeter, the elevation of each point must be established at least twice from independent runs.)

The right-hand page provides for entering the temperature adjustment (either by per cent formula or from tables), the preliminary adjusted elevation, the lineal adjustment, and the final elevations.

There are several methods used for altimeter leveling: stationary and roving altimeter, skip stop method, and double runs with base barograph. In all these methods the notes are arranged to allow the calculation to be included as part of the field procedure.

6-5 PROFILE NOTES

Profile notes record the elevation and description of points at short intervals along a fixed line which usually has been previously surveyed and staked.

Figure 63 illustrates a typical arrangement of profile notes using a Philadelphia rod.

The arrangement is much the same as that of level notes. The station numbers are entered in the proper column, and for clarity the profile elevations are shown in a column separate from the turning points. The descriptions of the elevation points are shown on the right-hand page. They should be kept as brief as clarity permits.

The use of ditto marks can save considerable writing.

Figure 64 shows notes for the same profile illustrated in Figure 63, using a Lenker rod instead of a Philadelphia rod. The form and arrangement are almost identical, but considerable recording time is saved by obtaining elevations directly from the rod.

Figure 65 illustrates profile notes using an instrument with a Beaman arc.

The left-hand page is a modified arrangement of a differential level form, and the Beaman arc data is recorded on the right-hand page.

The foresight rod appears to the right of the Beaman arc data, and the total difference in elevation is placed in the column usually used in the differential leveling form for the foresight rod. The difference in elevation value is accompanied by its proper algebraic sign, with the plus and minus values recorded on opposite sides of the column for clarity.

The form is interrupted horizontally by recording the station number of the instrument setup, the backsight rod, and the H.I. elevation, thus setting off each setup.

The problem here is to arrange the column headings in a manner that minimizes the amount of skipping around when recording, and still to maintain a note form similar to one with which the user is familiar.

6-6 CROSS SECTIONS

Cross-section notes record the elevations and descriptions of points referenced to a coordinate system selected for its compatibility with the proposed design calculations. For example, on a road design survey cross sections are taken at some simple subdivision of full stations plus appropriate breaks. They are taken at right angles or radially to the proposed alignment because this is compatible with the method used for volume calculations.

Figure 66 is an example of notes for a route cross-section survey. The arrangement of the left-hand page is similar to a differential level form except that it reads from bottom to top and the station numbers increase going up the page. Thus left and right are properly oriented when you are facing ahead on the line.

The right-hand page is divided by a vertical line representing the center line, and the individual rod readings are recorded over the distance out.

The turning point elevations are reduced in the field, but the cross-section points are not. They are left for the office to reduce.

Figure 67 shows cross-section notes kept in the profile form.

This arrangement requires the use of more pages than does the previous example (Figure 66) but is frequently more convenient, especially when it is inconvenient to have the rodman start each side of the section at the center line.

The example records station numbers increasing from the top to the bottom of the page, but the form can easily be reversed if it is advantageous.

In surveying rivers, streams, or creeks it is necessary to note the orientation of the lefts and rights. This is shown in the example at the top of the left-hand page. Actually this information would probably be placed in the general notes at the beginning of the work.

Observe the note at Station 56+68.4. This is an example of a cross reference to work not yet done.

Figure 68 shows cross-section notes in graphic form. The sketch is drawn approximately to scale. The positions of the shots are indicated by dimensioning and station numbering.

The spacing of the sections along Main Street through the center portion of the intersection are established by graphic construction, by holding the intersections of the theoretical flow line of the east gutter of Main Street and the center line and 10-foot offset lines of First Street. The actual station numbers can be calculated at a later time.

The use of a Lenker rod greatly reduced the amount of recording necessary.

The areas of concrete are shown cross-hatched. (Actually the use of colored pencils saves time and makes a clearer presentation.)

Figure 69 shows cross-section notes in descrip-

tive form using a sketch to clarify the grid coding method.

The form of the left-hand page is similar to the profile form with a column used to record the grid code number. The sketch on the right-hand page establishes the orientation and dimensions of the grid pattern.

Figure 70 shows cross-section notes using a Rhodes arc. In this arrangement the first column on the right-hand page is used to record the center line data, and the cross-section points are arranged to the left and right. The first column on the left-hand page is used for the station number. Both pages read from bottom to top.

At the center line, the height of the instrument is noted and the elevation of the center line stake is copied from the appropriate level book, or sometimes the center line elevation is filled in later by the office. The difference in elevation with the proper sign and the horizontal distance out are recorded directly from the arc. The terrain notation is abbreviated and appears below the arc data.

The title page in this case would carry notes stating that the instrument height is always sighted on the rod. When this is not done the rod reading must also be recorded. The title page should carry a proper reference to the source of the center line elevations.

Figure 71 illustrates notes for tunnel cross sections. In this type of survey the instrument is set up in the plane of the section, and the offset to tunnel center line is noted. The height of the spad above the instrument is recorded, with the elevation and source of the spad noted.

The instrument is set at 90° to the tunnel center line, and the vertical angles and distances to points on the tunnel wall are recorded. The vertical angles are usually laid off in even angular increments plus breaks.

Special care must be taken to indicate clearly the sign of the quadrant since most vertical circles are numbered by quadrants. The direction faced must be noted to avoid plotting the section in a mirrored position.

The sketch need only approximate the shape of the section, for it is intended to be used only as an aid in plotting.

It is not necessary to compute the actual elevations of the cross-section points because the office prepares the scale plot by polar coordinates and the area is determined by planimeter.

6-7 SLOPE STAKING

In slope staking notes the several trials to determine the catch point need not be recorded. Only data concerning the final stake need appear in the notes.

Figure 72 shows slope stake notes in the descriptive form. In this example the left-hand page is arranged like level notes, with the added column for the finished grade of the center line. The right-hand page is arranged to record on the center of the page the rod reading, the elevation of the center line stake, and the cut or fill at the center line.

The width of the road bed is noted at the top of the page, for example, left 25′.

After the catch point is determined and the slope stake set, the rod reading, elevation, cut or fill, and distance from center line are entered in the notes. When a reference stake is set, the rod reading, elevation, cut or fill, and offset also are entered in the notes. The slope is noted at the extreme left and right sides of the page.

The title page should contain appropriate notations regarding the type of stakes set and the type and marking of guard stakes, since these are usually constant for the project.

Figure 73 illustrates slope stake notes in the graphic form. This form is usually used for more complicated sections and requires preparing a sketch to show the proposed section and the existing ground line. Shown also are the control elevations, distances, and slopes for the proposed section.

The instrument H.I.'s are plotted on the sketch for clarity, and the cut or fill, distance out, and slope are noted for each point set.

This example assumes that the contractor requested that stakes be set at the wall face produced and the 1:1 slope produced to aid in staging the grading.

Note that the level circuit is shown taking off and closing on separate bench marks. If there is not enough space, this information should be placed on a separate page and cross-referenced.

Observe that the design stationing reads from the bottom of the page and the measured station and rods read from the top of the page. This was done to avoid confusing the two. In addition note the check distance to the tract line showing that the top of the cut is within tract limits.

Figure 74 illustrates a method of marking guard stakes for slope staking.

The back of the guard stake should always be marked with the station number of the section, should indicate left or right, and should be lettered

to read from the top of the stake down. The front of the stake should indicate whether it refers to a slope stake or a reference point. The slope stake shows the cut or fill, distance out, and the slope. The reference point shows the cut or fill and offset to the slope stake, then repeats the same information shown on the slope stake.

It is sometimes convenient to mark the front of the stake to read from the top down instead of across the stake.

When the design section is complicated grade sheets should be supplied, with sketches clarifying the relations of the stakes set to the design section.

Your field notes should always show the data marked on the guard stakes.

6-8 TOPOGRAPHY

Field notes for topographic surveys are arranged to provide space for both the field-measured values and the reductions.

Figure 75 shows a note arrangement for transit stadia notes for topography. The left-hand page column headings are arranged to record the stadia interval, horizontal angle, vertical angle, and rod reading. The horizontal distance, when computed, is entered on the right-hand side of the first column, and the difference in elevation and computed elevation is entered in the last two columns.

The description of the point read is entered on the left side of the right page, and the remainder of the page is used for a clarifying sketch.

Each setup is preceded by a notation of the point occupied, the direction the angles were turned, and the H.I. elevation for the setup.

The horizontal zero initial is recorded as the first reading of the set to establish the orientation.

Note the use of coding to show which building corners were read (three were taken to allow a check on the plotting).

Transit stadia notes involving any location always require a clarifying sketch.

In this example the reduction is left for the office to do.

Figure 76 shows notes for plane table topography. The columns are arranged to allow the readings and calculated values to be entered from left to right in the order in which they occurred, to achieve fast recording and reduction.

Speed in plane table topography is obtained through the teamwork of the observer and the recorder. The sequence of the observations, recordings, and computations must be well organized for maximum speed.

Each setup is preceded by the notation of the station occupied, backsight rod, and H.I. The page should be cross-referenced to the plane table sheet to which it applies in the event it later becomes necessary to check or verify any of the points read. The horizontal arc is not usually recorded because the horizontal distance is determined and plotted by the instrument man.

Figure 77 illustrates topography notes for strip topography kept in a combined cross-section and sketch form. In this example the route alignment has been previously staked and stationed.

The left-hand page is a combination of route cross-section and differential level notes reading up the page for orientation.

The right-hand page is a sketch to assist the user in correctly translating the notes into a topographic map.

The sketch shows the approximate interpolation of the contours, drainage lines, etc. Should the route alignment be on a curve, the radial sections are still plotted at 90° to the center line of the page. The sketch will be distorted, but distortion is acceptable here because the sketch is intended to be used only as an aid in interpreting the cross section. This is an exception to the rule that sketches should be drawn to an approximate scale.

6-9 CONSTRUCTION NOTES

Construction notes essentially consist of showing the location of stakes set for the control and construction of structures or grading. They frequently require the recording of a combination of angles, distances, leveling, and calculations to substantiate the fact that the work was laid out correctly.

Figure 78 shows construction notes for the stakeout of a culvert. The left-hand page records the differential levels necessary to determine the cut and fill, with a column for the design grades shown on the construction drawings.

The right-hand page contains a plan view of the culvert layout showing the horizontal position and relation of the stakes set. Because the sketch does not allow sufficient space to show how the guard stakes were marked, the information has been coded to the sketch and placed on the left-hand page. The rod readings are also coded for clarity.

Note that all distances and angles taken from the plans are marked "plans & meas." The title page will give a complete description of the plans and the pages used as well as any instructions concerning the required number of stakes and offsets requested by the contractor.

Figure 79 illustrates notes for the stakeout of a sewer. As in Figure 78, these notes are a combination of a sketch and differential level notes.

Since the marking of guard stakes is quite similar for each point set, the title page will contain a statement of the information shown on guard stakes and indicate which columns in the notes reflect the data shown on the stakes.

Figure 80 shows notes for the horizontal position of curb stakes, in graphic form. In this example a sketch is used for the street intersection to show the relation of the center lines, proposed curb lines, stake lines, and the details of laying out the curb returns.

It is not necessary to show graphically each 25′ station on the tangents between intersections since these are covered by the two notes shown.

Observe that the measured data for the control Stations 69+36.81, 70+11.21, and 3+59.33 are shown with the check-out distances. The check mark by each indicates that the check-out distances are within tolerances for the type of work being done. The curve data for both the curb line and the stake lines are shown.

Note that the existing paving along Fulton Avenue is shown, and the fact that Varna Avenue is not paved is noted as general information.

The stakes shown at Station 3+50 on Varna Avenue are noted as being previously set with a proper reference. The full notation is much clearer than simply calling them found stakes.

The title page will contain data about the type of stakes and guard stakes set, with a notation that each guard stake was marked with the station number, offset, and any other identifying marks such as P.V.C. (point on vertical curve). The same data will be repeated in the level notes along with the cuts and fills.

Figure 81 illustrates construction notes for slope staking an earth fill dam. In this example the procedure is described in full and accompanied by a sketch to explain and clarify further.

The method is excellent for accomplishing a difficult and time-consuming job, but since it is somewhat unusual particular care must be taken to explain the procedure in detail.

Observe that the method of checking is described; thus anyone reviewing the work will have confidence in the results even though he is unfamiliar with the method.

6-10 AS-BUILT NOTES

Notes of a survey to determine the as-built dimensions of structures are usually arranged in the graphic form showing three views, plan, profile, and cross section, along with enough special sections and details to depict the structure correctly.

It is usually not necessary to give all the small details you would expect to find on a set of construction drawings as long as enough principal dimensions are shown to supply the person using the notes with the information needed for his intended use.

The party chief should spend time with the user before making an as-built survey to determine exactly what information is required. For example, detailing a bridge for hydraulic studies usually requires detailing only below the bottom of the bridge stringer, whereas an as-built of a bridge for a proposed widening design is considerably more involved and requires the designer to supply the party chief with detailed, comprehensive instructions.

Figures 82 and 83 show as-built notes for a railroad girder bridge. This example assumes that the information obtained is sufficient to allow a designer to prepare plans to widen the bridge to a double-track width. It is also assumed that only the principal dimensions and elevations are required, since the designer himself will make field studies to determine the condition and structural details.

The notes are arranged to show a plan view and an elevation on which all the principal information is given. A cross section is shown to further clarify the plan and elevation view.

The abutments are given in plan and section views. (The two abutments are similar, and only one must be shown with the notation $\pm 0.2'$.)

Observe that the girder deflection was determined and noted on the profile view.

Certain elevations have been taken from the level notes (FB 12 pgs. 32–36) and shown on the profile for clarity.

In this example the title page will contain any general notes required to explain or clarify the survey.

Figures 84 and 85 illustrate as-built notes for a structure section through a levee. The assumption is that the levee is to be rebuilt. Therefore any structures passing through or across the levee need to be detailed with enough information that a de-

signer can rebuild or relocate the structure if necessary.

The notes are arranged to show plan and cross-section views with an enlarged detail of the pump house indicating the type of equipment it contains. The elevations here were obtained with a Lenker rod and recorded directly on the sketches in the proper location. Observe that these notes are cross-referenced to the proper books for the levels and soundings.

When making as-built surveys, often the party chief will be instructed to make an inventory of certain items. For example, on Figure 85 the equipment inside the pump house is identified, and it is only necessary to indicate its approximate shape and location with the appropriate notes regarding the trade name, size, and rating.

Whenever possible, it is preferable to make as-built surveys using a plane table, for such a procedure allows the sketches to be prepared to a larger scale and simplifies the showing of details.

6-11 BOUNDARY SURVEY NOTES

Boundary notes are almost invariably arranged in graphic form to show the relation of record lines, measured lines, found and set monuments, and other pertinent information.

The combined notes not only should tell the user what angles and distances you measured but should also clearly convey to him all the field information available so that the location can be verified in conformity to the theory of majority probability. The fact that you have established and monumented a corner in conformity to one principal of location—record, prorate, acceptance of acquiescent monuments, holding lines—does not eliminate the need to show all found monuments, natural and artificial, and all holding lines that are in conformity or in conflict with your location.

In property boundary notes, a most important objective is to record all facts of evidence of possession and encroachments. Equally important is the recording of the authority for acceptance of control points used as having zero positional error.

Figure 86 illustrates a note arrangement for recording the establishment of a working center line.

The lines established on this page, with those for the rest of the block, will be used as a basis for accomplishing a lot boundary survey.

It is not necessary to record the several trials required to finalize the center line. Only after the center line is fixed are the final distances measured and recorded.

All the evidence found is noted, and the substantiating reference to record is given when appropriate.

It is obvious from a study of the page that the center line of Fairview was established as the mean center line of the majority of the found points, giving weight to the found boat spike at the center line intersection of Fairview and Fifth Streets.

Observe that the curbs have been located and considered in the center line establishment. (They were probably constructed to stakes of some survey.)

Figures 87, 88, 89, and 90 show notes for a boundary survey. Figure 87 is arranged to show the record relation of the two tracts, assuming no conflict. Observe that the evidence of possession is noted, that is, the brick building and the stone wall.

As the survey progressed, and certain measurements and found monuments were noted, it became obvious that a conflict existed. Therefore the notes shown as Figure 88 were prepared to clarify and dimension the conflict.

In exaggerating the overlap, Figures 87 and 88, the scale relationship of the bisectors became confusing. Therefore the notes shown as Figure 89 were prepared for further clarification.

Observe the explanatory notes on all three figures, clearly stating the method of locating the west line of Tract 1265. The title page should contain the references to tract maps, center line tie notes, etc.

Field calculations are not covered in this handbook other than to state that, when made, they become a part of the notes and, therefore, are subject to the same rules of neatness, legibility, etc. Figure 90 shows the field calculations for the boundary survey illustrated in Figures 87, 88, and 89.

6-12 TRAVERSE NOTES

As previously mentioned, it is usually better to keep angles and distances on separate pages in the field book. It is sometimes necessary, however, to combine them, and when this is done care must be exercised in planning the note arrangement to be sure it fits the survey condition. There are literally dozens of arrangements that can be made for traverse notes, and it is very easy to select one and later in the survey find that it no longer fits the situation.

Figure 91 shows a note arrangement for a traverse using deflection angles and chained distances. The first four column headings on the left-hand page are arranged to allow the recording of point number, station number, deflection angle, and horizontal distance. The last two columns are reserved for recording slope distance and vertical angle. When a slope measurement is made the horizontal distance can be computed and entered in the horizontal distance column to allow the distance between points to be totaled.

The right-hand page shows a sketch with the page center line, representing the traverse line, disregarding the angle points. The direction of the deflection is indicated at each angle point by a dashed line in the approximate relation to the back tangent produced. When running deflection angle traverses it is good practice to record magnetic bearings as a check. These are recorded on the extreme left of the right-hand page.

Both pages read from bottom to top for orientation.

Figure 92 shows a note form for a stadia traverse carrying stadia elevations.

The arrangement reads from the top of the page to the bottom. The columns are arranged to record the point occupied, stadia interval, horizontal angle, vertical angle, rod reading, elevation, and point sighted.

The page is interrupted vertically by recording, for each point occupied, the height of the instrument above the station and the computed H.I. elevation of the station. The magnetic bearings are recorded on the left-hand edge of the page.

In this procedure the H.I. rod value for the station occupied is read on the foresight and backsight whenever possible, thus simplifying the computations. By such a procedure and by reading each distance twice (read both directions from each P.I.), the vertical angles and stadia distances can be field-compared as a check.

In this example the notes are reduced as an office procedure.

The method requires using two rodmen to achieve speed.

Figure 93 illustrates notes for a subtense bar traverse. The instrument is initialed on the backsight, and the angles are turned direct and inverted to each end and the center of the subtense bar. The note form is similar to that of triangulation notes using a T-2 theodolite (see Figure 51), with columns for reducing the angle across the bar and entering the distance taken from tables.

Each station occupied is recorded on a separate pair of pages, and sufficient positions are taken to insure the degree of accuracy required.

The description of the station mark is entered either below the angles or on the right-hand page.

Figure 94 illustrates notes for a subtense base traverse. The note arrangement is also similar to triangulation notes using a T-2 theodolite (see Figure 51), with the addition of a sketch showing the relation of the traverse stations and the eccentrics. The distance from the station being occupied to the eccentric is also given. (At least two distances are taken at different vertical angles.)

In this method the stations and the eccentrics are usually set by a reconnaissance party.

6-13 ASTRONOMIC OBSERVATIONS

In selecting a note form for recording astronomic observation, it is important to choose an arrangement that will allow fast recording of the points, for the time interval between direct and reverse pointings on the star is critical. For night work the form should not be crowded since it is difficult to record by artificial light.

Figure 95 shows a note arrangement for recording readings for an azimuth determination from solar observations, reading both verniers. The columns are arranged to record readings in the order in which they become available. Each set of readings is separated by leaving a line blank. The right-hand page should contain a sketch to clarify the notes and should also contain any support data such as the watch check.

Figure 96 illustrates notes recording the observations necessary for an azimuth determination by observing Polaris at any hour angle.

The column headings on the left-hand page are similar to triangulation notes with an added column for recording the time. The striding level readings are recorded on the left side of the right-hand page opposite the Polaris pointing. A sketch is shown on the right-hand page to clarify the orientation of the work.

Observe the note stating that the chronometer is set to read sidereal time; this is unusual but is shown to illustrate the importance of recording complete information.

When a striding level is used the notes must contain the value for one division of the level bubble.

6-14 HYDROGRAPHIC SURVEY NOTES

In hydrographic surveys one of the notekeeping problems is that of recording data for fixing the position of the launch as the soundings are taken. The form must allow rapid recording, since the fixes are taken at short time intervals to a moving target.

Figure 97 illustrates notes for cuts to a launch moving along range lines. The angles are read only once with the fix number and time being noted.

In this example it is assumed that when the observer arrived at the station he discovered that the line of sight to the backsight was obstructed (see Note 2). The data shown on the upper portion of the left-hand page is recorded to establish an eccentric setup to overcome the obstruction. The sketch further clarifies the eccentric.

Observe that the eccentric distance was measured twice, breaking chain as a check measurement to establish confidence in the work done by an inexperienced chainman (a fisherman, see Note 2).

In this example the observations are recorded starting at the bottom half of the left-hand page and continuing on subsequent pages. The columns are arranged to record fix number, direction, and time. (Times are always recorded on hydrographic work because they are a numerical check on the fix number and because the tidal stage correction is a function of the time.)

Observe the notes concerning watch synchronization, direction the angles were turned, and source of mark signal (radio) and the description of the specific point tracked on the launch. These are all necessary for comprehensive notes.

Figure 98 shows notes for launch fixes using two sextants. In this example all members of the party are in the launch. The position of the launch is plotted by the party chief on a circular plotting chart as the work progresses, and the notes are kept to allow a dry plot to be constructed in the office as the final work sheet.

Observe that all the support data is recorded and cross-referenced; the procedure used to accomplish the survey is explained in Notes 1 to 4.

The columns are arranged to record the fix number, both sextant readings, and the time, with space reserved for notations.

The sketch clarifies the orientation of the angles and makes clear the relation of the alternate angles observed by sextant "B." (For the first ten sextant "B" observes the angle to Station "Stack" and switches to Station M-21 when the sight to "Stack" becomes obstructed.)

Observe that the condition of the sea is given in addition to the weather note. (The sea condition has a direct effect on the expected accuracy of this type of survey.)

PART 7: summary

The preparation of a set of field notes requires considerable planning because of their importance to the successful completion of the survey.

Before starting to work the party chief must anticipate the type and amount of information to be recorded and exercise considerable judgment in selecting a suitable note arrangement.

He must bear in mind that his notes will be interpreted by others and must always submit notes that are complete, neat, legible, clear, and self-explanatory.

Honest notes show the pertinent information, measurements, and observations as of the date of the survey. They are never influenced by preconceptions and are always arranged to allow a check on the main scheme of the work.

Proper indexing and subdivision by title pages, page headings, and adequate references and cross references improve the continuity of field notes and add to their professional appearance.

Maximum use should be made of abbreviations, symbols, and codes to save time and space. Proper application of symbols and codes keeps the notes from becoming cluttered and results in improved clarity. Codes often serve another function, indicating the order in which the work was done.

Monument descriptions should be detailed and should clearly answer the questions *where* and *what*. When describing boundary monuments, particular emphasis is placed on recording data and observations that will assist in substantiating the record authority and legal acceptance of found monuments.

The recording of support data is necessary to establish the confidence level of the work. Support data may take many forms, from a photograph or a brief statement to a lengthy description of a procedure, observation, and conclusion.

The advisability of recording support data is a matter of judgment. The party chief must decide whether or not recording the data is necessary to establish an acceptable confidence level in the work.

In the final analysis the arrangement and preparation of field notes is a matter of experience and judgment, and it is very unlikely that two surveyors will prepare identical notes of the same survey.

The experienced party chief, understanding the fundamentals of notekeeping and having at his command a number of convenient note forms, combines, modifies, and adapts these to fit the particular conditions of a survey. By so doing he can design a note presentation containing all the necessary information, one that will permit only the correct interpretation of the notes.

There is today a strong tendency among surveyors to anticipate incorrectly the effect on notekeeping of future technical advances in automation. Automatic data recording and readout are being rapidly developed, and the acceptance and use by surveyors of new recording techniques are only a matter of time. Surveyors can undoubtedly look forward in the near future to attaching to their instruments such aids as analog takeoff devices which will record measurements automatically in a form acceptable by electronic computers.

Devices such as these will be valuable adjuncts to notekeeping and will have some effect on changing the form of the art, but it is certainly incorrect to assume that they will render the art obsolete. In fact, notekeeping will assume even greater importance as tools become more sophisticated. To maintain continuity in a set of notes that are composed of field books, graphs, magnetic tapes, punched cards, and so on, will require a great amount of skill, organization, and ability if the notes are to be interpreted correctly by both man and machine.

Regardless of technological advancements, the basic concept of field notes will remain unchanged; the future surveyor, like the ancient Egyptian, will be judged by his notes. It matters little whether they are recorded on magnetic tapes or clay tablets.

BIBLIOGRAPHY

1. Army, Department of, *Topographic Symbols.* Washington: Government Printing Office, 1961.

2. Bouchard, Harry, and Francis H. Moffitt, *Surveying.* 4th Edition. Scranton: International Textbook Co., 1959.

3. Breed, Charles B., and George L. Hosmer, *Elementary Surveying.* New York: John Wiley and Sons, 1951.

4. Brown, Curtis M., *Boundary Control and Legal Principles.* New York: John Wiley and Sons, 1957.

5. Davis, Raymond E., and Kelley Joe Wallace, *Short Course in Surveying.* New York: McGraw-Hill Book Co., 1942.

6. Gilbert, Basin W., "Leica Remembers When the Surveyor Forgets," *Leica Photography.* Fall, 1954, Vol. 7, No. 3.

7. Gossett, F. R., *Manual of Geodetic Triangulations.* C & GS S.P. 247, Washington: Government Printing Office, 1959.

8. Hoelscher, Randolph P., and Clifford H. Springer, *Engineering Drawing and Geometry.* New York: John Wiley and Sons, 1961.

9. Judson, Lewis U., *Calibration of Line Standards of Length and Measuring Tapes at The National Bureau of Standards.* NBS Monograph 15, Washington: Government Printing Office, 1960.

10. Kissam, Philip, *Surveying for Civil Engineers.* New York: McGraw-Hill Book Co., 1956.

11. Low, Julian W., *Plane Table Mapping.* New York: Harper and Brothers, 1952.

12. Nassau, Jason J., *Practical Astronomy.* New York: McGraw-Hill Book Co., 1948.

13. Pickels, George W., and Carroll C. Wiley, *Route Surveying.* New York: John Wiley and Sons, 1930.

14. Rice, Paul P., "Field Notes," *Surveying and Mapping.* Vol. IV, No. 3, pgs. 20–24, Washington: Universal Congress on Surveying and Mapping, 1944.

15. Surveying and Mapping Division, *Definitions of Surveyings, Mapping and Related Terms.* Manual 34, New York: American Society of Civil Engineers, 1954.

16. Surveying and Mapping Division, *Technical Procedure for City Surveys.* Manual 10, New York: American Society of Civil Engineers, 1957.

17. Swanson, L. W., *Topographic Manual.* C & GS S.P. 249, Washington: Government Printing Office, 1949.

18. Tracy, John C., *Surveying, Theory and Practice.* New York: John Wiley and Sons, 1953.

index of figures

APPENDIX: abbreviations

A.	Area	C.F.	Curb face
Abut.	Abutment	Ch.	Chain
Ac.	Acres	Chan.	Channel
A.C.B.	Asphalt concrete base	Chd.	Chord
A.C.Pav.	Asphaltic concrete pavement	Ch. X	Chiseled cross
A.C.W.S.	Asphalt concrete wearing surface	C.I.P.	Cast iron pipe or corrugated iron pipe
Adj.	Adjusted		
A.M.	Morning	Ck.	Check
Ang.	Angle	C/L or ¢	Center line
App.	Approximate	C.M.P.	Corrugated metal pipe
Asph.	Asphalt	Co.	County
Astro.	Astronomical	Conc.	Concrete
Av.	Average	Const.	Construction
Ave.	Avenue	Cont.	Continued
Az.	Azimuth	Cor.	Corner
Az. Mk.	Azimuth mark	Corr.	Corrected
& or ¢	And	Cos.	Cosine
		Cosec.	Cosecant
B.C.	Beginning of curve	Cot.	Cotangent
Bdry.	Boundary	C.R.	Curb return
Bk.	Book or bank	Ct.	Court
B.L.	Building line	C.T.	Copper tack
Bldg.	Building	Ctr.	Center
Blk.	Block	Ctr. Ret.	Center return
Blvd.	Boulevard	Cu.	Cubic
B.M.	Bench mark	Culv.	Culvert
Bot.	Bottom	C.W.S.	County wall sheet
B.S.	Backsight		
Btwn.	Between	D.	Degree of curvature
B.V.C.	Beginning vertical curve	Decl.	Declination
		Def.	Deflection
C.	Cut	Deg.	Degree
Calc.	Calculate	Descp.	Description
Cb.	Curb	Dest.	Destroyed
C.B.	Catch basin	D.F.	Douglas fir
C.C.P.	Centrifugal concrete pipe	D.G.	Disintegrated granite
C.E.	City engineer	D.H.	Drill hole
C.E.F.B.	City engineer's field book	Dia.	Diameter
Cem.	Cement	Diff.	Difference

Dist.	District or distance		J.C.	Junction chamber
D.M.	District map		J.S.	Junction structure
D.M.H.	Drop manhole			
Dr.	Drive		L.	Length of arc
Dwg.	Drawing		Lat.	Latitude
			L.B.	Left bank
E.	East		Lb.	Pound
E.C.	End of curve		L. Chd.	Long chord
e.g.	For example		L.D.	Local depression
El. or Elev.	Elevation		L.&T.	Lead and tack
Ellip.	Elliptical		L.H.	Lamp hole
Engr.	Engineer(ing)		Lin. Ft.	Lineal foot or feet
E/O	East of		Lks.	Links
Eq.	Equation		Loc.	Locate or location
E.R.	End of return		Log.	Logarithm
Esmt.	Easement		Long.	Longitude or longitudinal
Estab.	Established		L.S.	Land surveyor
Etc.	Etcetera (and so forth)		Lt.	Left
E.V.C.	End vertical curve		L.W.	Low water
Ex.	Existing			
Ext.	External		M.	Meter
			Max.	Maximum
F.	Fill		M.B.	Map book
F.B.	Field book		M.C.	Middle of curve
F.C.	Flood control		Meas.	Measure
Fd.	Found		Mer.	Meridian
F.H.	Fire hydrant		Mi.	Mile
Fig.	Figure		Mic.	Micrometer
F.L.	Flow line		Mid.	Midway or middle
F.S.	Foresight or finished surface		Mil. Res.	Military reservation
Ft.	Foot or feet		Min.	Minimum or minute
F.T.	Flush tank		Misc.	Miscellaneous
			M.H.	Manhole
G.	Gutter		M.H.H.W.	Mean higher high water
Galv.	Galvanized		M.H.T.	Mean high tide
G.C.	Grade change		Mk.	Mark
G.L.	Ground line		Mkd.	Marked
Gr.	Grade		M.L.L.W.	Mean lower low water
			mm.	Millimeter
H.C.	Head chainman or house connec-		M.O.	Middle ordinate
	tion		Mon.	Monument
H.I.	Height of instrument		Mp.	Map
Hor.	Horizontal		M.R.	Miscellaneous record
Hr.	Hour		M.S.L.	Mean sea level
Ht.	Height			
H.W.	High water		N.	North
Hy.	Highway		N.&T.	Nail and tin
			Nat.	Natural function
i.e.	That is		N.E.	Northeast
In.	Inch		No.	Number
Inst.	Instrument		N/O	North of
Inters.	Intersection		**Norm.**	Normal
I.P.	Iron pipe		N.W.	Northwest

Obs.	Observe		R.&O.	Rock and oil
O.D.	Outside diameter		R.P.	Reference point or radius point
Opp.	Opposite		R.R.	Railroad
O.R.	Official records		R.R.R/W	Railroad right of way
O.&W.	Opening and widening		Rt.	Right
Ord.	Ordinance		R/W	Right of way
Orig.	Original		Ry.	Railway
Pg.	Page		S.	South or sewer
Pgs.	Pages		San.	Sanitary
Par.	Parallel		S.B.B.&M.	San Bernardino Base & Meridian
Pav.	Paved, paving or pavement		S.B.M.D.	Standard bench mark disk
P.B.M.	Precise bench mark		S.D.	Storm drain
P.C.	Party chief		Sdg.	Sounding
P.C.C.	Point of compound curve		S.E.	Southeast
Perp.	Perpendicular		Sec.	Secant
P.I.	Point of intersection		Sect.	Section
P.I.Q.	Property in question		Sht.	Sheet
P.L. or ₤	Property line		Sin.	Sine
Pla.	Place		S/O	South of
P.L.P. or ₤ Prod.	Property line produced		S.&T.	Spike and tin
P.M.	Afternoon		S.&W.	Spike and washer
P.O.B.	Point of beginning		Spec.	Specifications
P.O.C.	Point on curve		Spec. pub.	Special publications
P.O.S.C.	Point on spiral curve		Spk.	Spike
P.O.S.T.	Point on semi-tangent		Sp. M.H.	Special manhole
P.O.T.	Point on tangent		Sq.	Square
P.P.	Power pole or picture point		Sq. Ft.	Square foot or feet
P.R.C.	Point of reverse curve		S.S.	Sanitary sewer
Prod.	Produced or product		S.S.M.	Standard survey monument
Prof.	Profile		St.	Street
Prop.	Proposed		S.T.	Semi-tangent
Pt.	Point		Sta.	Station
P.T.	Point of tangent		Std.	Standard
Pvt. R/W	Private right of way		Stk.	Stake
Pvmt.	Pavement		Stks.	Stakes
			S.T.M.	Standard traverse monument
			Str. Gr.	Straight grade
Rad.	Radial or radius		Struct.	Structure or structural
R.B.	Right bank		Surv.	Survey
R.C.	Rear chainman		S.W.	Southwest
R.C.P.	Reinforced concrete pipe			
Rd.	Road			
Rdy.	Roadway		Tan.	Tangent
Rec.	Record		T.B.M.	Temporary bench mark
Recon.	Reconnaissance		Tel.	Telephone (pole)
Ref.	Reference		Temp.	Temporary or temperature
Res.	Reservoir		Term.	Terminus
Ret. wall	Retaining wall		Terr.	Terrace
Rge.	Range		T.H.	Test hole
R.H.	Redhead		Tk.	Tack
R.M.	Reference mark		₤̄	Throw over
Ro.	Rancho		T.L.	Traverse line

Topo.	Topography	V.C.	Vertical curve
T.P.	Turning point	Vert.	Vertical
Tr.	Tract		
Trans.	Transition	W.	West
T.S.	Traffic signal	W.C.	Witness corner
Twp.	Township	Wk.	Walk
		W.L.	Water line
U.S.C.E.	United States Corps of Engineers	W/O	West of
U.S.C.&G.S.	United States Coast & Geodetic Survey	W.P.	Witness post
		W.S.	Wearing surface or water surface
U.S.G.S.	United States Geological Survey	X-sec.	Cross section

40

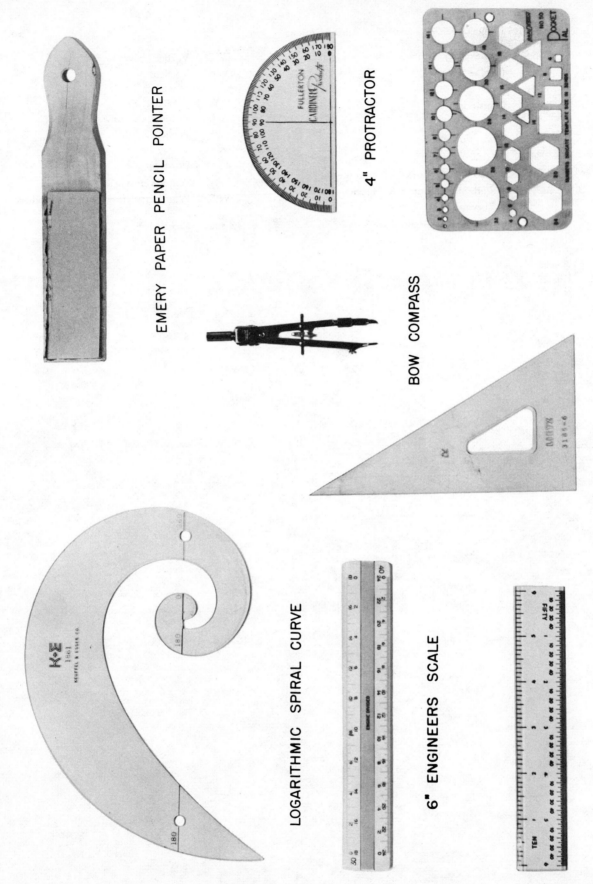

EMERY PAPER PENCIL POINTER

4" PROTRACTOR

PLASTIC TEMPLATE

BOW COMPASS

6" 30°/60° TRIANGLE

LOGARITHMIC SPIRAL CURVE

6" ENGINEERS SCALE

6" PLASTIC SCALE & STRAIGHT-EDGE

FIGURE I

UPPER CASE ARE CAPITAL LETTERS

ABCDEFGHIJKLMNOPQRSTUVWXYZ
1234567890 SLANT

SLOPE 2 TO 5

2/5

3/5

ABCDEFGHIJKLMNOPQRSTUVWXYZ
1234567890 VERTICAL

PRACTICE MAKING LETTERS LARGE AND
USE GUIDE LINES - YOUR LETTERING WILL LOOK
BETTER WHEN REDUCED

PRACTICE 15 MINS A DAY FOR 2 WEEKS AND
NOTE THE IMPROVEMENT

FIGURE 2-A

41

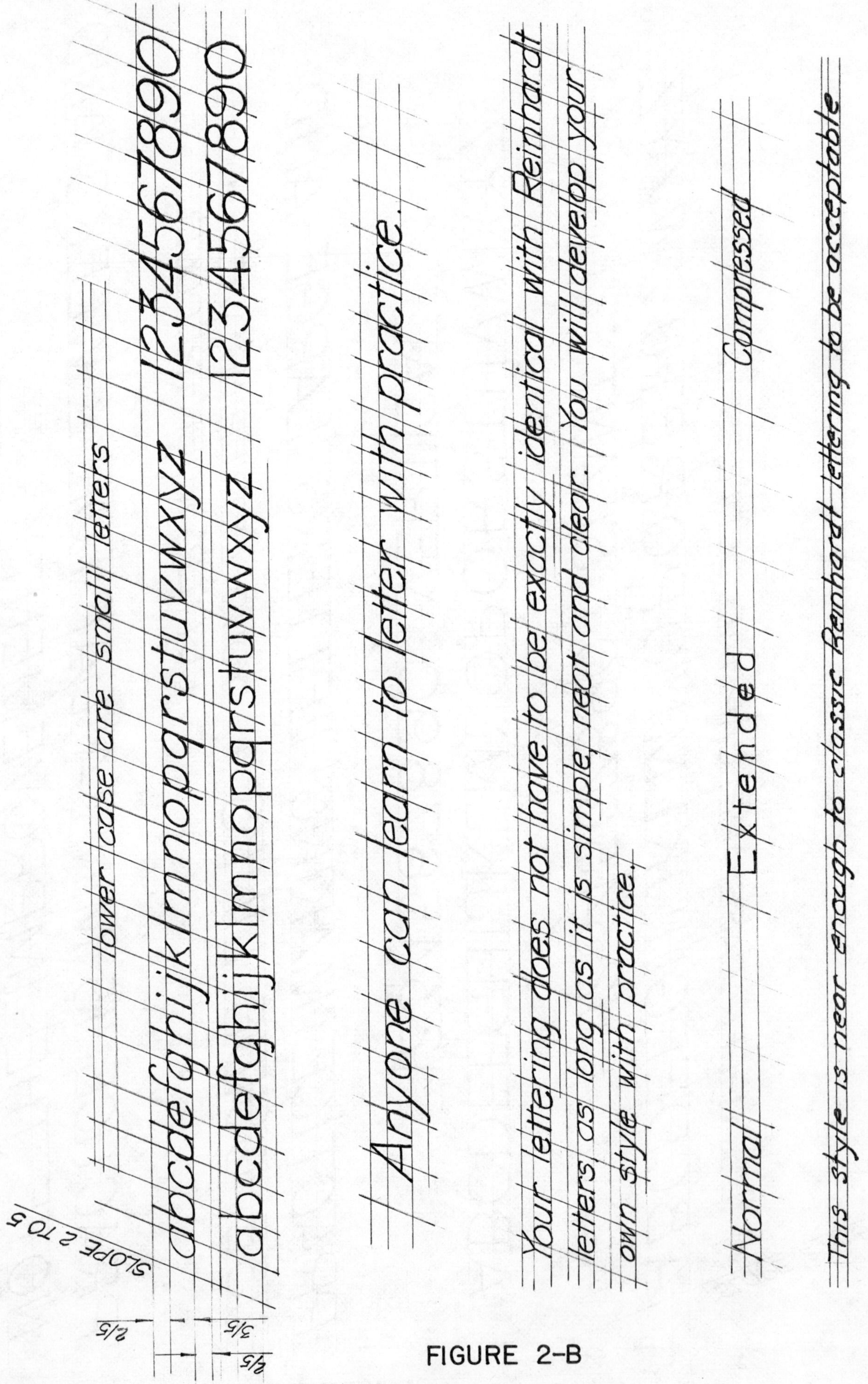

42

lower case are small letters

abcdefghijklmnopqrstuvwxyz 1234567890

abcdefghijklmnopqrstuvwxyz 1234567890

SLOPE 2 TO 5

2/5

3/5

2/5

Anyone can learn to letter with practice.

Your lettering does not have to be exactly identical with Reinhardt letters as long as it is simple, neat and clear. You will develop your own style with practice.

Normal

Compressed

Extended

This style is near enough to classic Reinhardt lettering to be acceptable

FIGURE 2-B

INDEX

Description	Pages	Job Number	Date	Party Chief
1) SAINT VINCENT CHURCH - Culvert stakes	5-9	5-1003	Nov 10 1957	A. Smith
2) PACIFIC SHIPYARD - Drop hammer Foundation stakes & detail	10-14	4-327	Dec 4 1957	B. Jones
3) REDONDO HYDRO - Control traverse	15-30	4-331	Jan 3 1958	A. Smith
4) LOT 10, TR 7334 - Bdry Survey	31-38	4-339	Feb 16 1958	C. White
5) FRESNO J.C. - Topographic Survey Control Traverse	39-55	4-343	Mar 5 1958	B. Jones
Control Bench Levels	56-65			
Transit Topo	66-78			
(Transit Topo Cont. in Book 21 Pg 34)				
6) TRACT 21,500 - Offsite sewer profile	79-84	4-349	June 20 1958	A. Smith

FIGURE 3

CHRONOLOGICAL INDEX

See Section 2-2

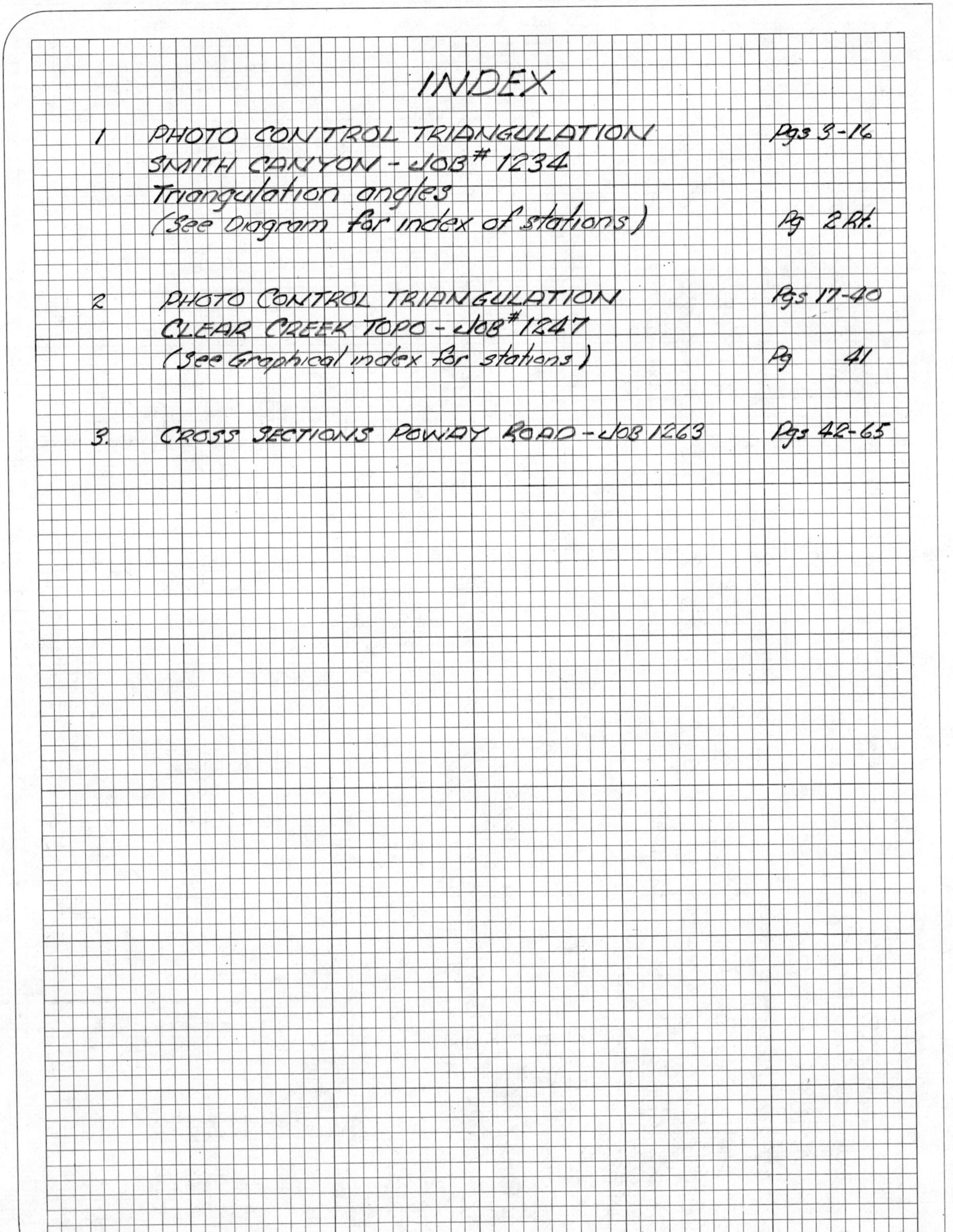

INDEX

1	PHOTO CONTROL TRIANGULATION SMITH CANYON - JOB # 1234 Triangulation angles (See Diagram for index of stations)	Pgs 3-16 Pg 2 Rt.
2	PHOTO CONTROL TRIANGULATION CLEAR CREEK TOPO - JOB # 1247 (See Graphical index for stations)	Pgs 17-40 Pg 41
3.	CROSS SECTIONS POWAY ROAD - JOB 1263	Pgs 42-65

FIGURE 4

CHRONOLOGICAL INDEX

See Section 2-2

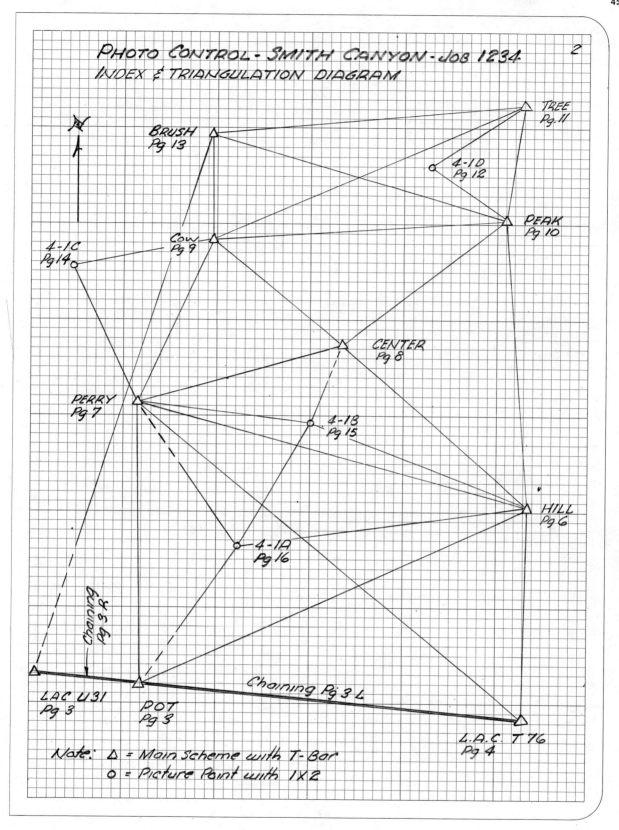

FIGURE 5

COMBINED INDEX AND TRIANGULATION DIAGRAM
See Section 2-2

46

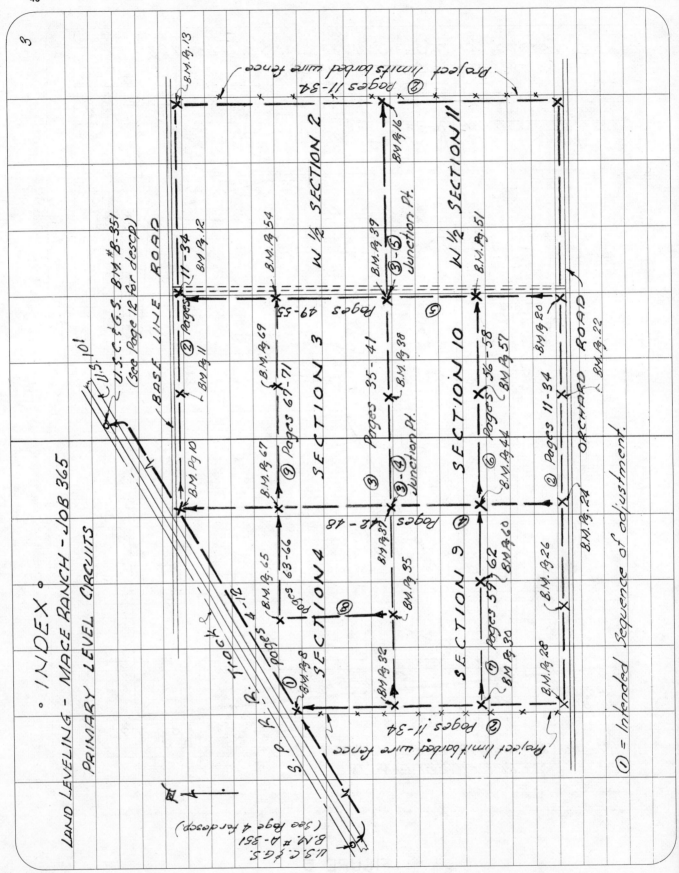

FIGURE 6

GRAPHICAL GEOGRAPHIC INDEX
See Section 2-2

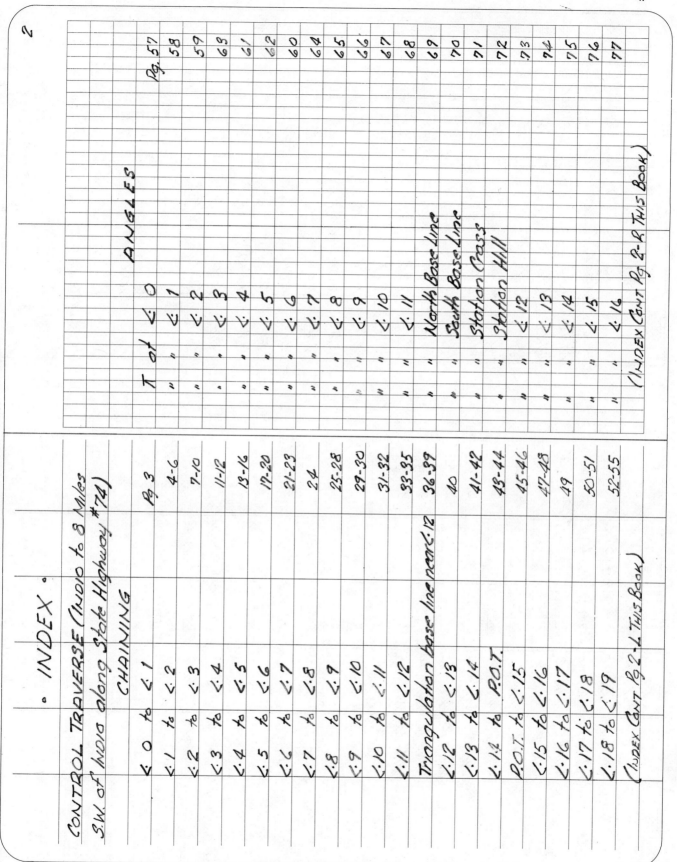

FIGURE 7

SPLIT BOOK INDEX IN NUMERICAL SEQUENCE
See Section 2-2

INDEX ° Job 5725

SOLINAS RIVER LEVELS
Level Run, West bank from Solinas to 40 Miles North

BENCH MARK DESCRIPTIONS

Description	Pages	Bench Mark Descriptions	Page
Mile 0 - Mile 2	Pg 3-10	Mile 1.15	Pg 5
Mile 2 - Mile 4	11-18	Mile 1.93	10
Mile 4 - Mile 6	19-22	Mile 3.10	15
Mile 6 - Mile 8	23-30	Mile 4.02	18
Mile 8 - Mile 10	31-40	Mile 5.12	20
Mile 10 - Mile 12	41-52	Mile 6.10	22
Mile 12 - Mile 14	53-64	Mile 7.02	25
Mile 14 - Mile 15.27	65-70	Mile 7.98	29
(Cont. in Level Book 7 Pg 6)		Mile 9.17	36
		Mile 10.10	41
		Mile 10.51 (Extra B.M.)	43
		Mile 11.02	47
		Mile 12.08	53
		Mile 12.89	59
		Mile 14.10	64
		Mile 15.03	70
		(Cont. in Level Book 7 Pg 6)	
SPUR LINE WEST FROM B.M. Mile 10.10			
Mile 10.10 - West 1.3 Miles	72-76	Mile 10.10 Aux #1	74
TRANSFER B.M. 14.10 TO HIGHER GROUND			
Mile 14.10 thru B.M Mile 14.10 Reset to B.M. Mile 15.03	78-79	Mile 14.10 Reset	79

3

48

FIGURE 8

DOUBLE INDEX IN ARBITRARY SEQUENCE
See Section 2-2

49

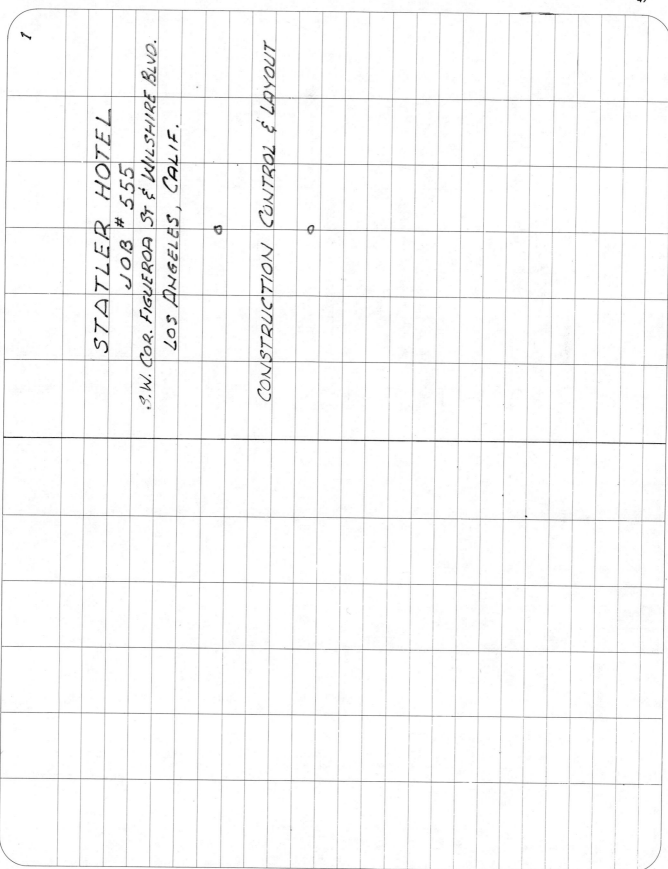

FIGURE 9

TITLE PAGE FOR A COMPLETE BOOK
See Section 2-3

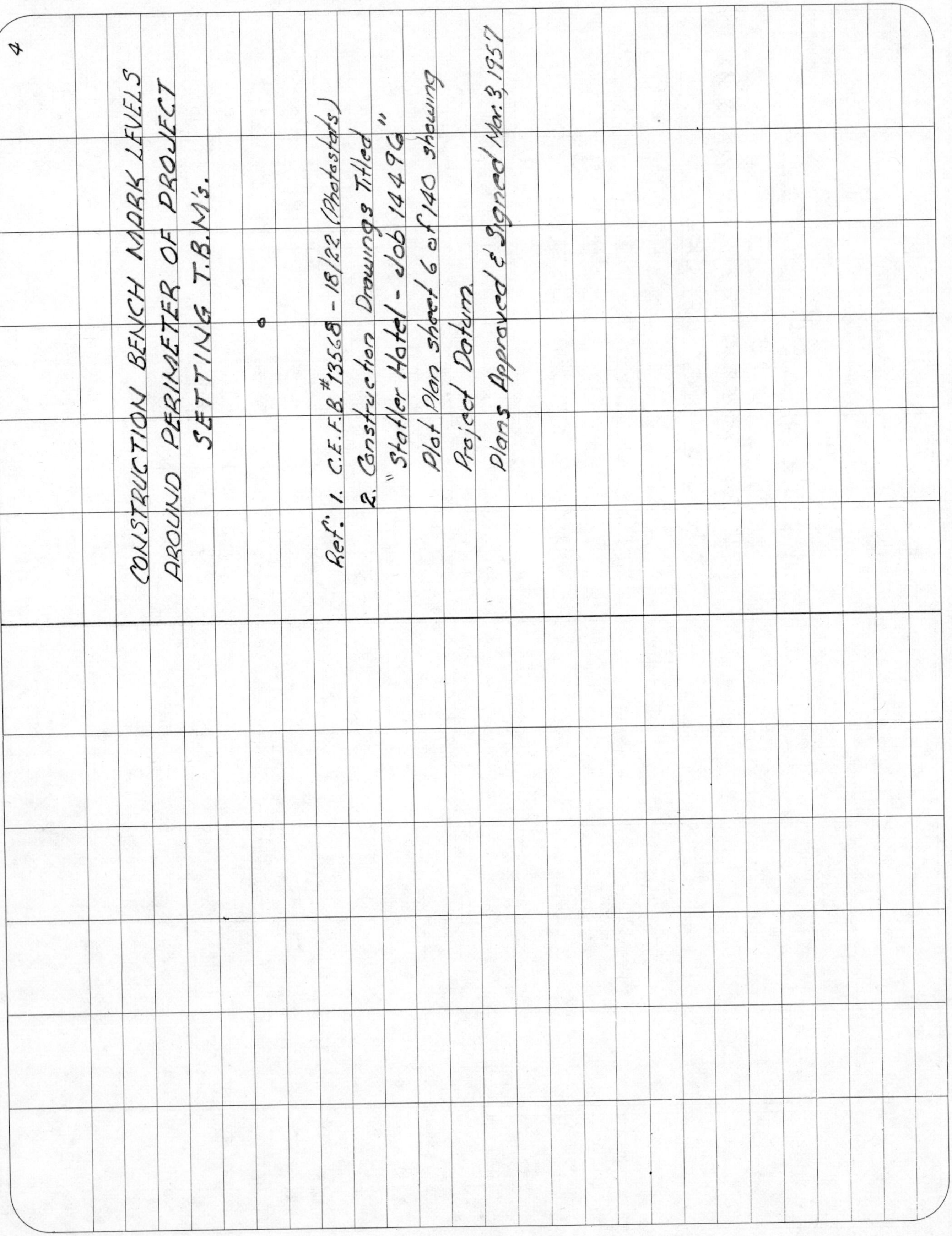

50

4

CONSTRUCTION BENCH MARK LEVELS
AROUND PERIMETER OF PROJECT
SETTING T.B.M's.

Ref. 1. C.E.F.B. #13568 - 18/22 (Photostats)
2. Construction Drawings Titled
"Statler Hotel - Job 1449G"
Plot Plan sheet 6 of 140 showing
Project Datum
Plans Approved & Signed Mar. 3, 1957

FIGURE 10

SUBTITLE PAGE FOR A COMPLETE BOOK
See Section 2-3

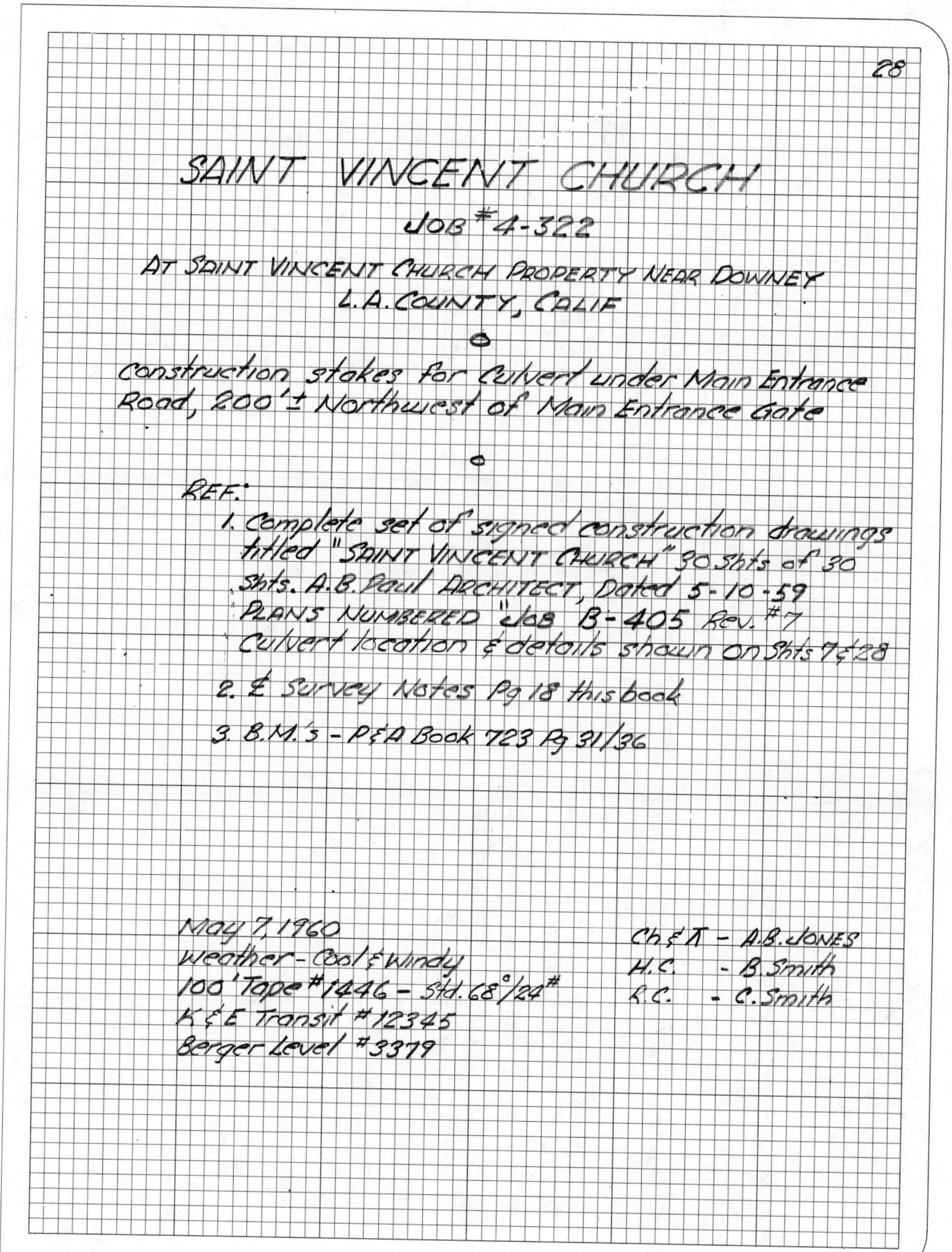

28

SAINT VINCENT CHURCH
JOB #4-322
AT SAINT VINCENT CHURCH PROPERTY NEAR DOWNEY
L.A. COUNTY, CALIF

⊕

Construction stakes for Culvert under Main Entrance
Road, 200'± Northwest of Main Entrance Gate

⊕

REF.:

1. Complete set of signed construction drawings
   titled "SAINT VINCENT CHURCH" 30 shts of 30
   Shts. A.B. Paul ARCHITECT, Dated 5-10-59
   PLANS NUMBERED "JOB B-405 Rev. #7
   Culvert location & details shown on Shts 7 & 28

2. ℄ Survey Notes Pg 18 this book

3. B.M.'s - P&A Book 723 Pg 31/36

May 7, 1960
Weather - Cool & Windy
100' Tape #1446 - Std. 68°/24#
K&E Transit #12345
Berger Level #3379

Ch & R - A.B. JONES
H.C. - B. Smith
R.C. - C. Smith

FIGURE 11

TITLE PAGE, JOB OF ONE-DAY DURATION
See Section 2-3

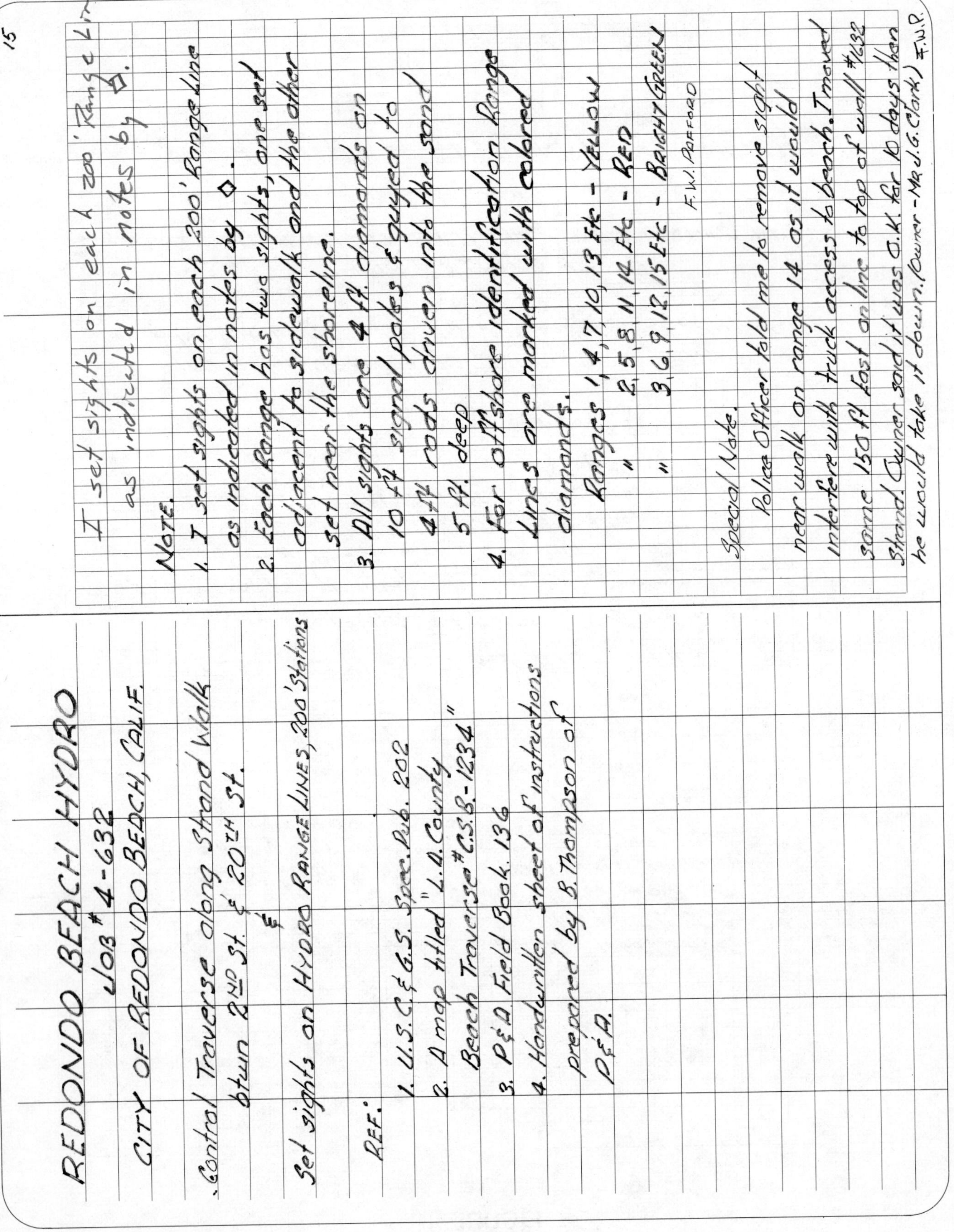

I set sights on each 200' Range Line as indicated in notes by ◇.

NOTE:

1. I set sights on each 200' Range Line as indicated in notes by ◇.

2. Each Range has two sights, one set adjacent to sidewalk and the other set near the shoreline.

3. All sights are 4 ft. diamonds on 10 ft. signal poles & guyed to 4"x4" roots driven into the sand 5 ft. deep.

4. For offshore identification Range Lines are marked with colored diamonds.

Ranges 1,4,7,10,13 ft - Yellow

" 2,5,8,11,14 ft - Red

" 3,6,9,12,15 ft - Bright Green

F. W. Pofford

Special Note.

Police Officer told me to remove sight near walk on range 14 as it would interfere with truck access to beach. I moved some 150 ft East on line to top of wall #632. Strand Owner said it was O.K. for 10 days then he would take it down. (Owner - Mr J.G. Clark) F.W.P.

REDONDO BEACH HYDRO

Job # 4-632

CITY OF REDONDO BEACH, CALIF.

Control Traverse along Strand Walk btwn 2nd St. & 20th St.

Set sights on Hydro Range Lines 200' Stations

REF:

1. U.S.C.& G.S. Spec. Pub. 202
2. A map titled "L.A. County Beach Traverse #C.S.B.-1234"
3. P&'D Field Book. 136
4. Handwritten sheet of instructions prepared by B. Thompson of P.&'D.

FIGURE 12

TITLE PAGE, JOB OF SEVERAL DAY'S DURATION

See Section 2-3

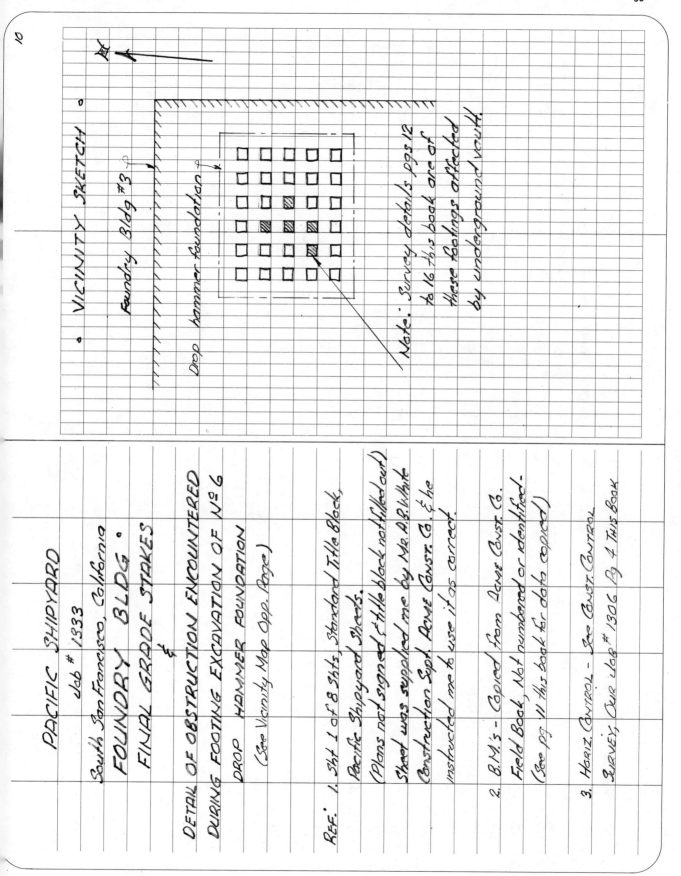

FIGURE 13

TITLE PAGE, COMBINED WITH A SKETCH
See Section 2-3

53

54

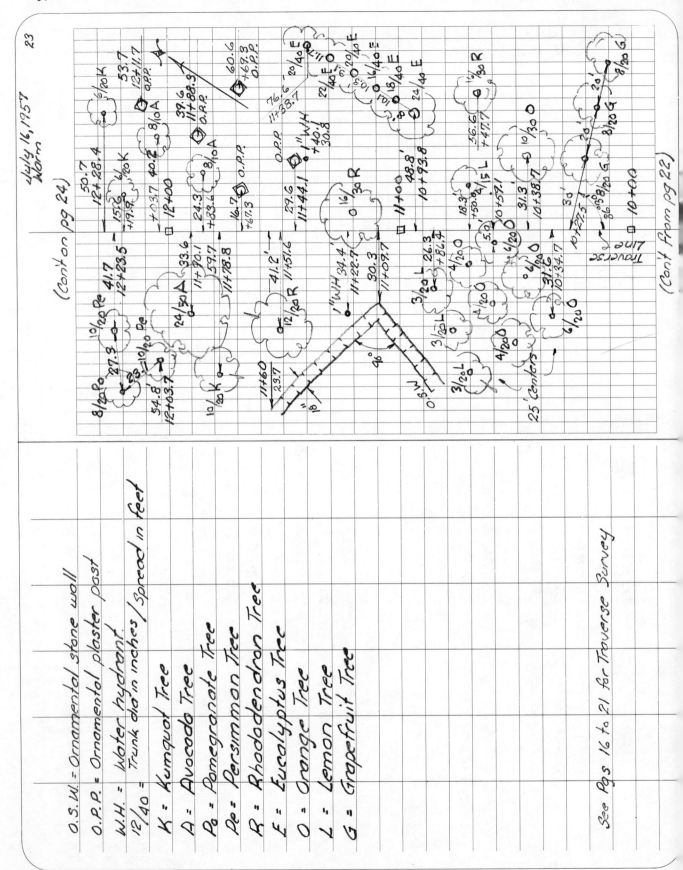

See Pgs 16 to 21 for Traverse Survey

O.S.W. = Ornamental stone wall
O.P.P. = Ornamental plaster post
W.H. = Water hydrant.
12/40 = Trunk dia'n inches / Spread in feet

K = Kumquot Tree
A = Avocado Tree
Po = Pomegranate Tree
Pe = Persimmon Tree
R = Rhododendron Tree
E = Eucalyptus Tree
O = Orange Tree
L = Lemon Tree
G = Grapefruit Tree

FIGURE 14

USE OF ABBREVIATIONS IN LOCATION NOTES
See Section 3-1

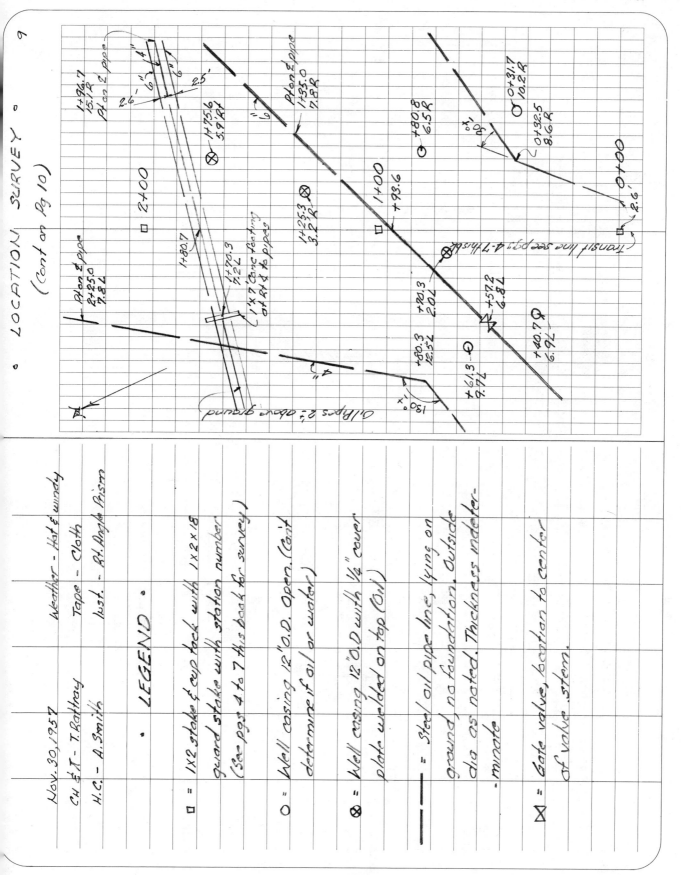

FIGURE 15

USE OF SYMBOLS IN LOCATION NOTES

See Section 3-2

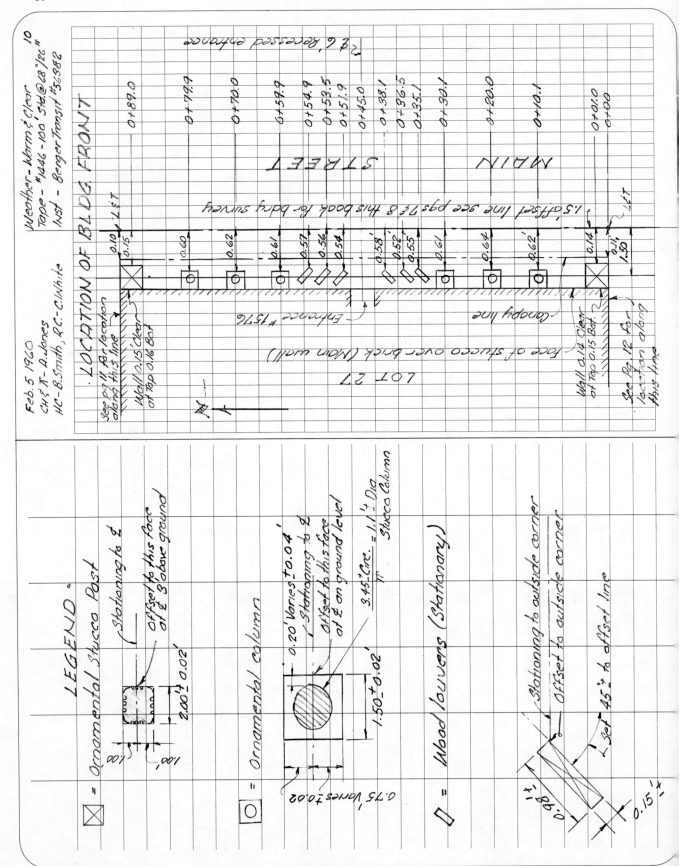

56

FIGURE 16

USE OF SYMBOLS IN LOCATION NOTES
See Section 3-2

//	Parallel to
⊥	Right angle
 ⫫	Same both sides of line (usually ownership)
@	At
—·—·—	Centerline or traverse line
=	Equal
≡	Identically equal to or exactly
≠	Is not equal
∼	Similar to
≅	Approximately equal
<	Is less than
>	Is greater than
≧	Equal to or greater than
≦	Equal to or less than
∴	Therefore
∵	Because
 ⌇	Intermittent stream
 ⨿⨿⨿	Stone fence
 ∧∧∧	Rail fence
 –×–×–×	Barbed wire fence
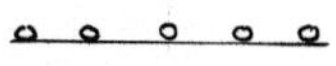 ○○○○○	Woven wire or chain link fence
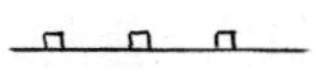 ⊓⊓⊓	Board fence
 ∿∿∿	Hedge

FIGURE 17

SYMBOLS
See Section 3-2

⊼	Instrument or instrument man
 ▱	Notekeeper or recorder
⸓	Rod or rodman
o	Iron pipe
□	Stake
-¦-	Planetable setup
⊡	Stadia station
⊙	Traverse station
△	Triangulation station
⊠	Bench mark
∠	Angle point or P.I.
⊿	Angle
#	Number or pound
△	Delta, central angle
△⁄2 △⁄4	$\frac{1}{2}$ delta, $\frac{1}{4}$ delta, etc.
ϕ	Phi, latitude
λ	Lambda, longitude
ϕ̄	Throw-over
π	Pi=3.14159265+
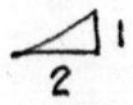 ⊿₂¹	Slope 2 to 1
Σ	Summation
⊥	Perpendicular to

FIGURE 17

SYMBOLS
See Section 3-2

	Evergreen tree
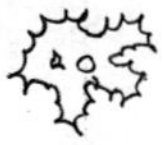	Deciduous tree
TP	Turning point
$\ominus ... \oplus$	Sign is reverse of normal
①, ②	Indicates 1st, 2nd, measurement of same quantity

FIGURE 17

SYMBOLS

See Section 3-2

60

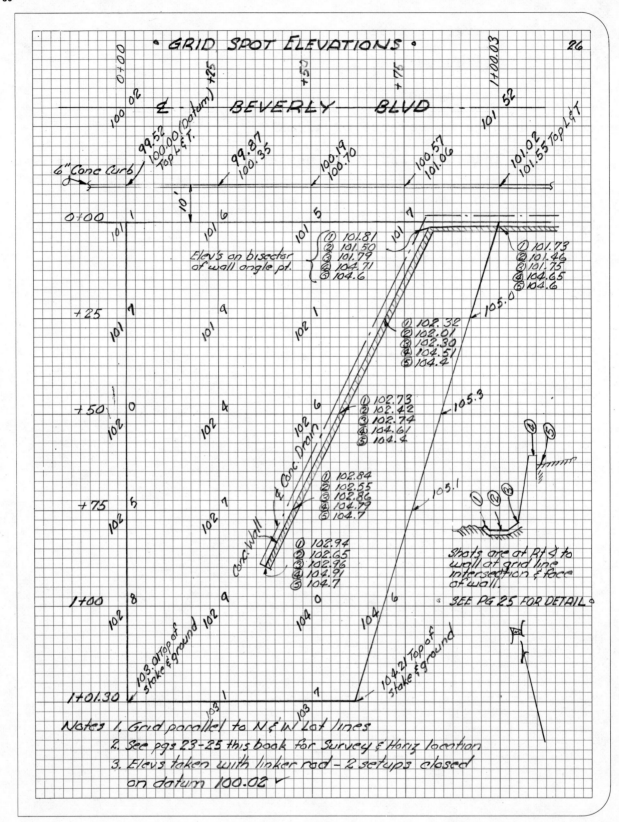

FIGURE 18

NUMERICAL CODING IN TOPOGRAPHIC NOTES
See Section 3-3

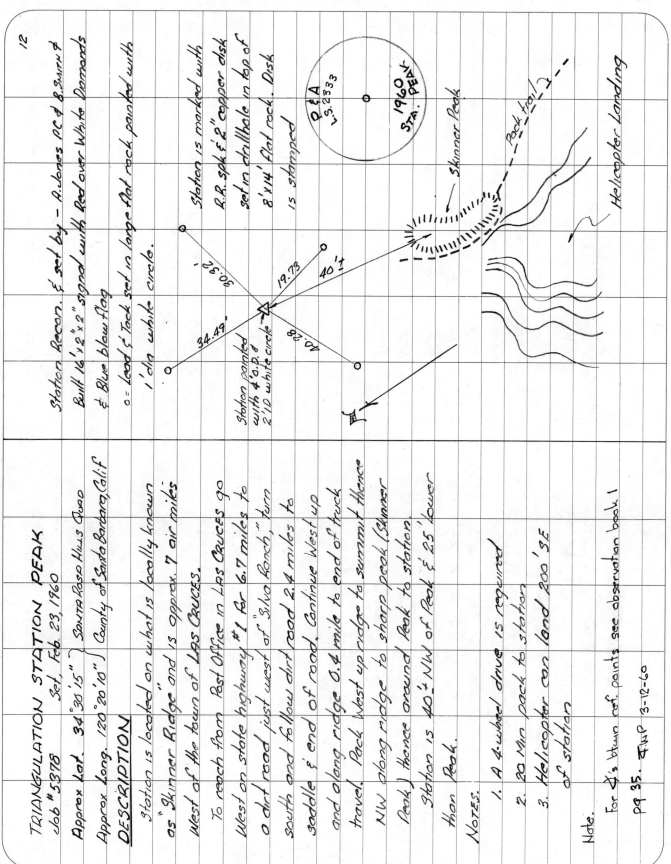

TRIANGULATION STATION PEAK
Job #5378   Set. Feb. 23, 1960
Approx. Lat. 34°30'15")   Santa Rosa Hills Quad
Approx. Long. 120°20'10") County of Santa Barbara, Calif.
DESCRIPTION

Station is located on what is locally known as "Skinner Ridge" and is approx. 7 air miles West of the town of LAS CRUCES.

To reach from Post Office in LAS CRUCES go West on state highway #1 for 6.7 miles to a dirt road just west of "Silva Ranch," turn south and follow dirt road 2.4 miles to saddle & end of road. Continue West up and along ridge 0.4 mile to end of truck travel. Pack West up ridge to summit thence NW along ridge to sharp peak (Skinner Peak) thence around Peak to station. Station is 40'± NW of Peak & 25' lower than Peak.

NOTES.
1. A 4-wheel drive is required
2. 20 Min pack to station
3. Helicopter can land 200' SE of station

Note.
For ±'s btwn ref. points see observation book 1
Pg 35. FWP 3-12-60

Station Recon. & set by — A. Jones Rc & B. Smith J
Built 16'± 2'x2" signal with Red over White Demands & Blue bloss flag

o = Lead & Tack set in large flat rock painted with 1 dia white circle.

Station is marked with R.R. spk & 2" copper disk set in drillhole in top of 8'x14' flat rock. Disk is stamped

Station painted with 4' a.d. & 2'ID white circle

P & A
L.S. 2333

1960
STA. PEAK

Skinner Peak

Pack trail

Helicopter Landing

19.73

30.32'

34.49'

40'±

40.28

12

61

FIGURE 19

TRIANGULATION STATION DESCRIPTION
See Section 4–4

13

NOTES, BENCH MARK DESCRIPTIONS

1. All Bench Marks are plotted on a copy of Yolo Quad submitted with these notes as a part of this survey.

2. All Bench Marks are Photo identified on photos supplied - See Ref. #1 for numbers

3. B.M.'s are monumented with a bronze disk 3" dia with a 3" shank set in a 8"x8"x3' conc. post set flush with ground. The disk is stamped thus

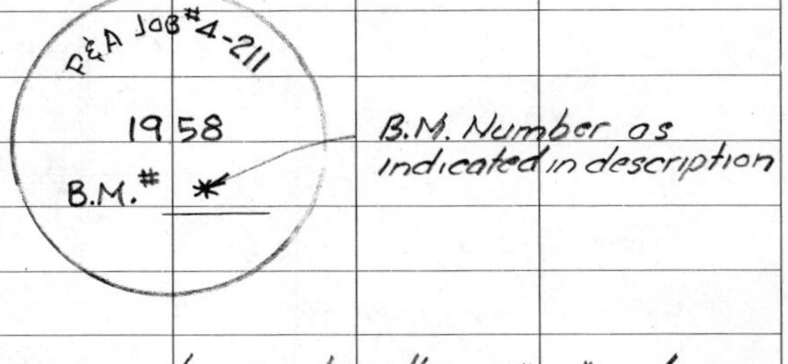

```
        R & A Job # 4-211

              1958

           B.M. #  ─*
```

B.M. Number as indicated in description

4. All B.M.'s are witnessed with a 4"x4"x5' redwood post painted white & set 3' into ground & offset as indicated in the description.

5. On the following pages when a description refers to a portion of a previous description and indicated thus (Same as 3-17) it refers to only that portion of the previous description that is underlined.

YOLO MAPPING
Job 4-211
T.12N, R.7E, M.D.B.&M.
Yolo County, Calif.

THIRD ORDER LEVELS
Along south side of T.12N, R.7E, M.D.B.M.
Establishing B.M. & Picture Point Elevations

REF.
1. Photos 2-4, 2-6, 2-8, 2-10, 2-12, 2-14, 2-16, 2-18, 2-20, 2-22, 2-24 & 2-26 - Job 4-211
2. U.S.C.&G.S. 1st Order Leveling Spr. Rob #281 "Sacramento to Walnut Grove, Calif."
3. Sketch and written data concerning location of B.M.'s prepared by A. White

Crew - D.B. Jones
  " - B. Smith
  " - B. Perry
Wild Level #13356 - N-3
K&E Semi Precise Rods #5 USC&G 51568
Rod targets used & rods alternated on T.P.'s

FIGURE 20

LEVEL NOTES, SINGLE WIRE – ALTERNATE RODS
BENCH MARK ACCUMULATIVE DESCRIPTION
See Sections 4-4 and 6-4

TWO PEG TEST

	Far Rod		Near Rod	
Dist.	Reading		Reading	Dist
	TP "B"		TP "A"	
① 298	3.797		4.064	32
	TP "A"		TP "B"	
② 300	4.737		4.466	34
578	8.534		8.530	66
−66	−0.004*		8.530	
532	8.530		0.000 ✓	

* Corr. for curvature & refraction

LEVEL CHECKS FLAT.

Sights	B.S.	H.I.	F.S.	Elev	Adj.Elev	Dist		
						B.S.	F.S.	
C 351	1.732	93.935		92.223	92.223	210		U.S.C.&G.S. B.M. C-351. Recovered monument in good condition as per description in Spec. Pub. 281 Sht 10.
TP	2.632	92.932	3.655	90.300		280	220	
TP	6.341	93.462	5.811	87.121		280	290	
TP	10.180	101.575	2.067	91.395		250	270	
TP	6.328	104.846	3.057	98.518		260	250	
TP			3.465	101.381 ✓			210	
	27.213		18.055			1280	1290	

+ 27.213
− 18.055
+ 9.158
92.223
101.381 ✓

7-16-58
Cool & Clear

14

(Cont on Pg 15)

FIGURE 21

LEVEL NOTES, SINGLE WIRE – ALTERNATE RODS
BENCH MARK ACCUMULATIVE DESCRIPTION
See Sections 4–4 and 6–4

Sights	B.S.	H.I.	F.S.	Elev.	Adj. Elev.	Dist. B.S.	Dist. F.S.	Description	Date
		(Cont. from Pg 14)							7-16-58 Cool & Clear
B.M. 3-17	1.975	103.356		101.381		190		B.M. #3-17* Commencing at John Marshall High	
TP	3.622	103.982	2.996	100.360		210	190	School in town of Yolo thence S.W. along	
TP	8.344	107.896	4.480	99.552		320	270	County Road 5.6 Mi. to "T" road, turn NW and	
TP	12.180	119.009	1.067	106.829		275	310	Continue along dirt road 2.1 Miles; B.M. set	
TP	8.285 / −8.385	125.219	2.075	116.934		340	275	21'N of ℄ of road & 10'W of irrigation head-	
P.P. 4-C	3.080	124.834	3.465	121.754		240	350	gate. Witness Post set 3.5' North	
TP	1.575	114.369	12.040	112.794		290	240	Photo Point 4-C - 1X2 stake (See back of photo	
TP	5.648	107.547	12.470	101.877		200	300	2-4 for description & identification)	
B.M. 3-18	7.442	107.956	7.033	100.514		230	190	B.M. 3-18* (Same as 3-17) 4.7 Mi.; B.M.15 set	
TP	5.739	107.185	6.510	101.446		220	240	137'N of fence, 254'W of NW cor of stucco	
TP	7.137	110.376	3.946	103.239		230	210	house, 35'NE of 30" Black Walnut Tree.	
TP			11.271	99.105			230	Witness Post set 27' East.	
	65.027		67.303			2795	2805		

(Cont. on Pg 16)

−67.303
+65.027
−2.276
101.381
−2.276
99.105 ✓

FIGURE 22

**LEVEL NOTES, SINGLE WIRE—ALTERNATE RODS
BENCH MARK ACCUMULATIVE DESCRIPTION**
See Section 4-4 and 6-4

16
7-16-58
Cool & Clear

Sights	B.S.	H.I.	F.S.	Elev	Adj. Elev	Dist. B.S.	F.S.	Description
		(Cont. from Pg. 15.)		99.105				
P.P. 4-D	2.705	101.810		99.105			210	PHOTO POINT 4-D - 1X2 stake (See back of photo 2-4 for description & identification)
	2.362	100.503	3.669	98.141		250	210	
TP	7.434	103.593	4.344	96.159		310	260	
B.M. 3-19	11.178	113.164	1.607	101.986		280	300	B.M. #3-19* (same as 3-17) 5.1 Mi. to gravel cross road, turn SW and continue 1.1 Mi.;
	9.282	119.871	2.575	110.589		310	280	B.M. set 150 NW of £ road, 70 SW of windmill. Witness Post set 3.1' SW.
TP	⊕ 1.832	119.495	⊕ 1.456	121.327		240	320	Rod inverted on nail in power pole
	1.775	110.136	11.134	108.361		290	230	" " "
B.M. 3-20	6.584	106.016	10.704	99.432		200	280	B.M. #3-20* (same as 3-17 & 3-19) 2.8 Mi.; B.M. set in NW fence line 30' NW of fence cor. Witness Post set 2.5' N.W.
	8.242	107.955	6.303	99.713		210	190	
	5.739	108.084	5.610	102.345		210	220	
	7.713	112.833	2.964	105.120		220	210	
			10.712	102.121		230	220	
	61.182		58.166			2730 = 2730		
		(Cont. on Pg 17)						

+ 61.182
- 58.166
+ 3.016
99.105
102.121

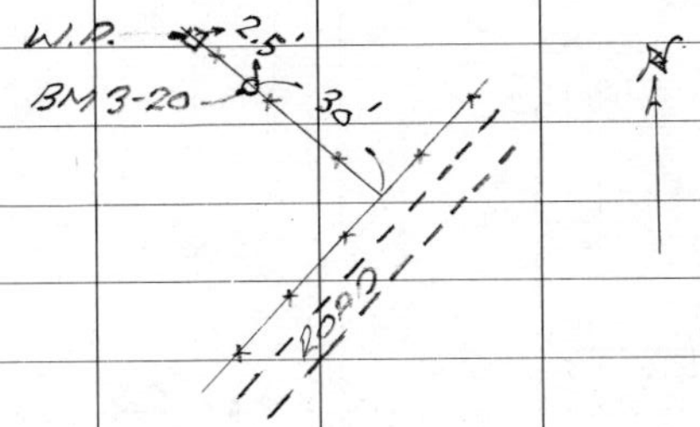

FIGURE 23

LEVEL NOTES, SINGLE WIRE – ALTERNATE RODS
BENCH MARK ACCUMULATIVE DESCRIPTION
See Sections 4-4 and 6-4

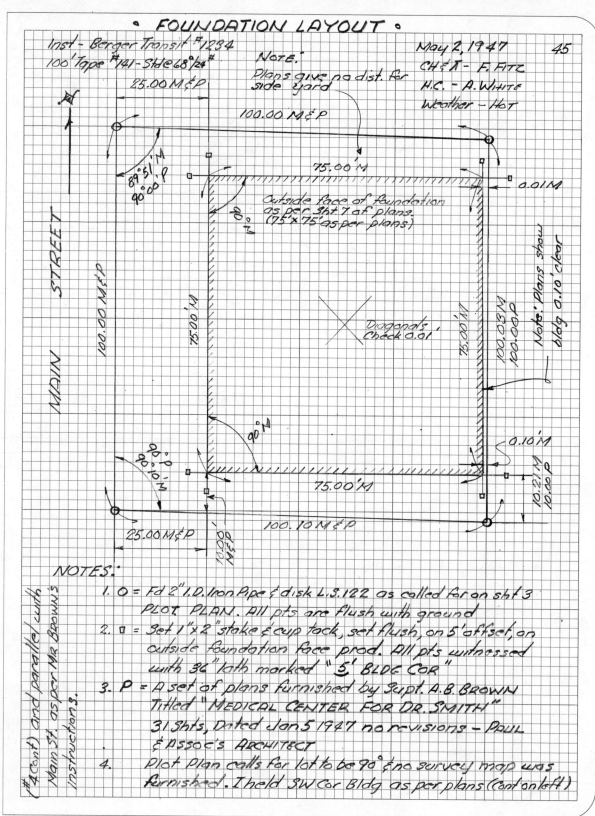

° FOUNDATION LAYOUT °

Inst - Berger Transit #1234
100' Tape #141 - Std 68°/24#

NOTE:
Plans give no dist. for
side yard

May 2, 1947   45
CH &T - F. FITZ
H.C. - A. WHITE
Weather - Hot

25.00 M&P
100.00 M&P
89°51'M
90°00'P
90°00'
0.01M

Outside face of foundation
as per sht 7 of plans
(75'x75' as per plans)

75.00'M

Note: Plans show
bldg 0.10 clear

MAIN   STREET
100.00 M&P
75.00'M
75.00'M
100.03M
100.00'P

Diagonals,
Check 0.01

90°M
90°P
90.10'M
75.00'M

0.10'M
10.21'M
10.00'P

25.00 M&P
100.00'M&P
100.10 M&P

NOTES:

1. O = Fd 2" I.D. Iron Pipe & disk L.S.122 as called for on sht 3
   PLOT PLAN. All pts are flush with ground

2. □ = Set 1"x2" stake & cup tack, set flush, on 5' offset, on
   outside foundation face prod. All pts witnessed
   with 36" lath marked "5' BLDG COR"

3. P = A set of plans furnished by Supt. A.B. BROWN
   Titled "MEDICAL CENTER FOR DR. SMITH"
   31 Shts, Dated Jan 5 1947 no revisions - PAUL
   & ASSOC's ARCHITECT

4. Plot Plan calls for lot to be 90° & no survey map was
   furnished. I held SW Cor Bldg as per plans (Cont on left)

(#4 Cont) and parallel with
Main St. as per Mr BROWN's
Instructions.

FIGURE 24

CONSTRUCTION MONUMENT DESCRIPTIONS
See Section 4-5

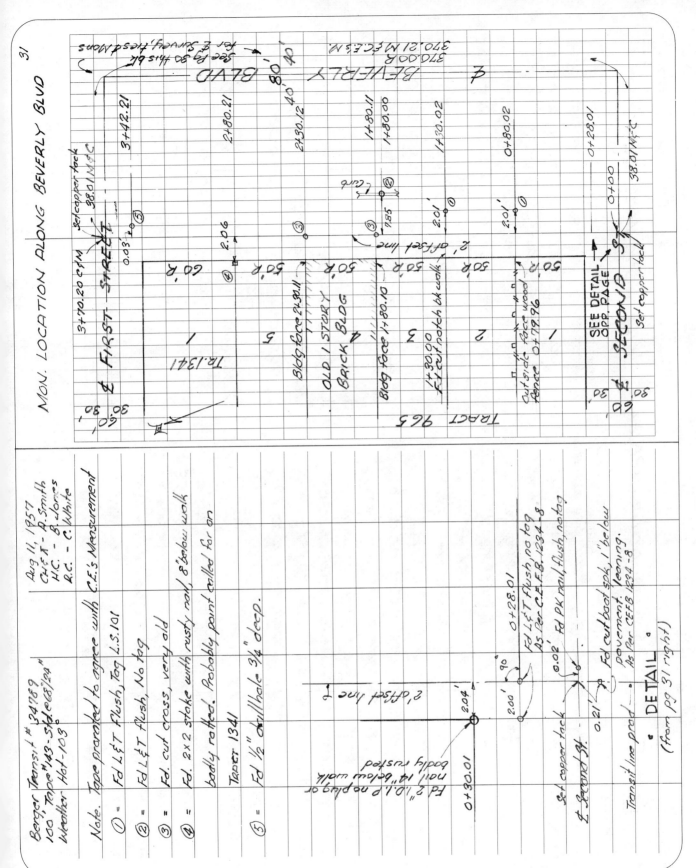

FIGURE 25

BOUNDARY MONUMENT DESCRIPTIONS

See Section 4-6

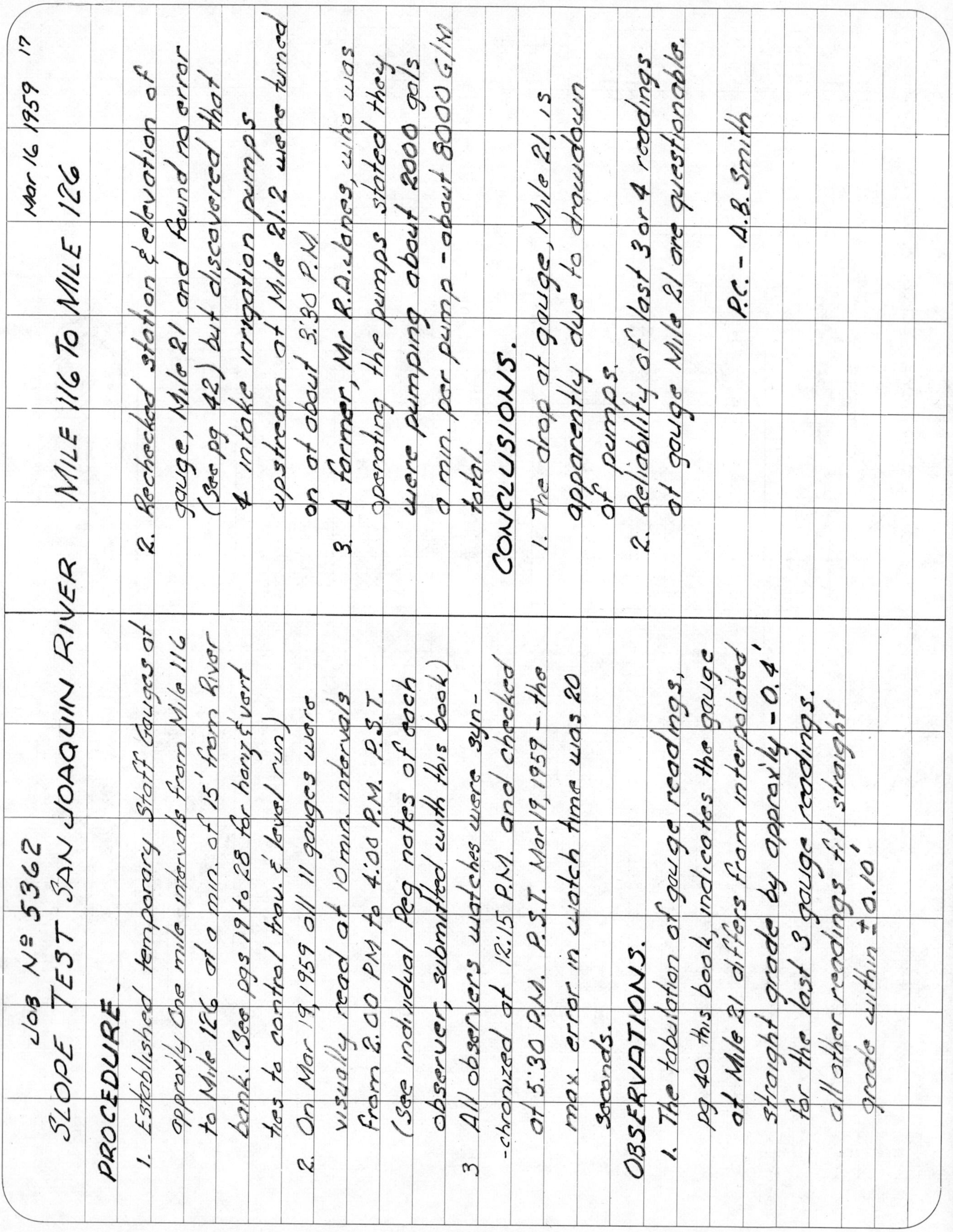

Job Nº 5362   Mar 16 1959   17

SLOPE TEST SAN JOAQUIN RIVER   MILE 116 TO MILE 126

PROCEDURE.

1. Established temporary Staff Gauges at approx'ly. One mile intervals from Mile 116 to Mile 126 at a min. of 15' from River bank. (See pgs 19 to 28 for horiz'l vert ties to control trav. & level run.)

2. On Mar 19, 1959 all 11 gauges were visually read at 10 min. intervals from 2.00 P.M. to 4.00 P.M. P.S.T. (See individual Peg notes of each observers, submitted with this book.)

3. All observers watches were syn-chronized at 12:15 P.M. and checked at 5:30 P.M. P.S.T. Mar 18, 1959 — the max. error in watch time was 20 seconds.

OBSERVATIONS.

1. The tabulation of gauge readings, pg 40 this book, indicates the gauge at Mile 21 differs from interpolated straight grade by approx'ly −0.4' for the last 3 gauge readings. All other readings fit straight grade within ± 0.10'

2. Rechecked station & elevation of gauge, Mile 21, and found no error (See pg 42) but discovered that 4 intake irrigation pumps upstream at Mile 21.2 were turned on at about 5:30 P.M.

3. A farmer, Mr R.D.Jones who was operating the pumps stated they were pumping about 2000 gals a min. per pump - about 8000 G/M total.

CONCLUSIONS.

1. The drop at gauge, Mile 21,15 apparently due to drawdown of pumps.

2. Reliability of last 3 or 4 readings at gauge Mile 21 are questionable.

P.C. − A.B. Smith

FIGURE 26

DESCRIPTION OF A PROCEDURE, OBSERVATION & A CONCLUSION
See Section 5-2

STATEMENT PURSUANT TO AUTHORITY UNDER
CHAPTER 15, DIVISION 3, ART. 5, SECTION 8760 OF
CALIFORNIA BUSINESS & PROFESSIONS CODE.

STATE OF CALIFORNIA } S.S.
COUNTY OF LOS ANGELES }

    I, John B. Doe, being first duly sworn, do depose
& state the following:
    I have often seen the government post at the North corner
common to Section 4 & 5, Township 4 North, Range 15
East, S.B.B.& M. and I have read the markings scribed
on said post identifying it as the cor of Sections 4 & 5.
    It was at my suggestion that my uncle S.C.Doe
drove an iron axle 3/4" in dia along side the post
as the post was becoming very decayed.
    The axle was placed there in the year 1904
or 1905.

John B Doe
JOHN B. DOE

    Subscribed & Sworn to before me, a Licensed
Surveyor, in and for said State, this 25th day
of Dec., 1936

I. Wm P.
Licensed Surveyor #2333

FIGURE 27

OATH, OBLITERATED CORNER
See Section 5-3

70

25

STATEMENT PURSUANT TO AUTHORITY UNDER CHAPTER 15, DIV. 3, ART 5, SECTION 8760 OF CALIFORNIA BUSINESS & PROFESSIONS CODE

STATE OF CALIFORNIA)
                    } S.S.
COUNTY OF SAN DIEGO)

We, A.B. Smith & V.J. Brown, being first duly sworn, do depose & state the following:

The decomposed remains of a redwood post, indicated as point "A" in the sketch below, was in existence in the year 1906, the date we jointly built the barbed wire fence between our properties and that we used said stake to control the location of said fence, further we have always accepted said post as the corner marked for the corner of Lots 5 & 6.

The redwood post indicated as point "B" in the sketch below was set by a surveyor, identity unknown to us, sometime around the years 1914 to 1915, and we witnessed him driving said post in the ground.

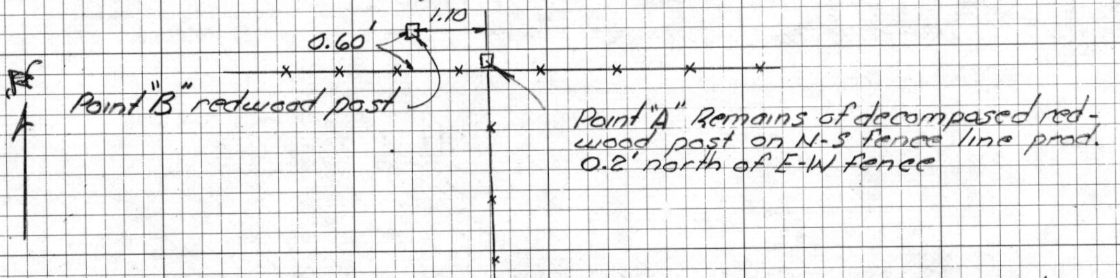

Point "B" redwood post

0.60'   1.10

Point "A" Remains of decomposed redwood post on N-S fence line prod.
0.2' north of E-W fence

We also observed MR F.Wm PAFFORD replace the post indicated as point "A" with a 2" I.D.I.P set in concrete, flush with the ground, with a nail & surveyors brass disk stamped L.S.2333. He did this March 9, 1960

A.B. Smith
A.B. SMITH

V.J. Brown
V.J. BROWN

Subscribed & sworn to before me, a licensed surveyor in and for said State, this 10th day of March, 1960

F.W. Pafford
Licensed Surveyor #2333

FIGURE 28

OATH, MONUMENT IN PERISHABLE CONDITION
See Section 5-3

57

STATEMENT PURSUANT TO AUTHORITY, UNDER CHAPTER
15, DIV. 3, ART. 5, SECTION 8760 OF CALIFORNIA BUSINESS &
PROFESSIONS CODE.
STATE OF CALIFORNIA ⎞
COUNTY OF LOS ANGELES ⎠ s.s.

    We, W.A. FITZ & J.M. PERRY, being first duly sworn, do
depose and state the following:
    That during the course of surveying the boundary
of that parcel of land known as Lot 10, Tract 12345,
M.B 21 Pg. 10/17, in the City of Pasadena, County of
Los Angeles, State of California, on Jan 20, 1953, that
we faithfully performed our duties as chainmen
with care and accuracy, and to the best of our
ability.
    Further, that upon completion of the survey
we remeasured all our original measurements
as a check and found them to agree within
± 0.01 foot.

_____
W.A. FITZ

_____
J.M. PERRY

    Subscribed and sworn to before me, a Licensed
Surveyor, in and for said State, this 20th day
of Jan., 1953

_____
F.W. PAFFORD
Licensed surveyor #2333

FIGURE 29
OATH, FAITHFUL PERFORMANCE OF DUTIES
See Section 5-3

TEST & ADJUST K&E DUMPY LEVEL #12345

(A) Adjust level vial
① Direct - Bubble centered
Reverse - Bubble 2½ grads left
Removed ½ error with capstan
② Direct - Bubble centered
Reverse - Bubble centered ✓ ok.

(B) Adjust horiz crosshair
① Check pt on hair at right field of view -
high at left side.
Rotated crosshair ring to adjust
② Repeat check - low at left side
Rotated crosshair ring to adjust.
③ Repeat check - ✓ O.K.

Jan. 24, 1948
Adjusted by - A. Smith
Hot & Dry

(C) Two Peg Test

Far Rod		Near Rod	
Dist.	Reading	Reading	Dist.
	T.P."A"	T.P."B"	
① 299	4.520	4.264	35
	T.P."B"	T.P."A"	
② 298	3.580	3.862	32
597	8.100	8.126	67
−67	−0.004*	8.096	
530	8.096	+0.030	

$+0.030 \div 5.30 = +0.0057$ "c" per 100 ft.
$+0.0057 \times 2.98 = 0.017"$ Raised line of sight this amount

	T.P."B"	T.P."A"	
① 298	3.797	4.064	32
	T.P."A"	T.P."B"	
② 300	4.737	4.466	34
598	8.534	8.530	66
−66	−0.004*	8.530	
532	8.530	0.000 ✓ O.K.	

* Corr. for curvature & refraction

(D) Rechecked vial & crosshair and found OK ✓

FIGURE 30

TEST & ADJUST DUMPY LEVEL
See Section 5-4

TEST & ADJUST GURLEY WYE LEVEL # 3756

April 6, 1949
Adjusted by - A.Smith
Cool & Clear

Ⓐ Adjust crosshairs

① Locked crosshairs and et on signboard, rotated scope 180° in collars and found crosshairs high & to right. Corrected ½ error with capston and on recheck crosshairs OK ✓

Ⓑ Adjust level vial

① Direct - Bubble centered
② Reversed - Bubble centered ✓ ok

Ⓒ Adjust Wyes

① Direct - Bubble centered
② Reversed Bubble 1½ grads right
Removed ½ error with wye capston
① Direct Bubble centered
② Reversed - Bubble centered ✓ ok

Ⓓ Two Peg Test

	Far Rod		Near Rod	
	Dist.	Reading	Reading	Dist.
		T.P."A	T.P."B"	
① 280		4.620	4.334	20
		T.P."B"	T.P."A"	
② 281		3.570	3.851	19
561		8.190	8.185	39
-39		-0.004*	8.186	
522		8.186	-0.001	

-0.001' ÷ 522 = -0.0002' = C" Per 100'

Inst In good adjustment ✓ ok

* Corr. for curvature & refraction

FIGURE 31

TEST & ADJUST WYE LEVEL

See Section 5-4

Oct. 19, 1951
Adjusted by - D. Jones
37

Adjusted inside warehouse
Cool, clear & quiet

TEST & ADJUST DIETZGEN #4476 TRANSIT

(A) Adjust plate bubbles
Transverse Bubble (1) Direct - Bubble centered
(2) Reversed - Bubble 1 grad left
Removed 1/2 error with capstan until Bubble checked ✓
Side Bubble (1) Direct - Bubble centered
(2) Reversed - Bubble centered ✓ OK

(B) Preliminary adjustment vertical hair
Top of hair on mark as scope elevated
hair is to left. Rotated crosshair ring until
hair stays on mark as scope elevated

(C) Adjust vertical hair to double center.
Double center at 600' sights (Error = 1.20')
Corrected 0.30' with capstans.
Repeated until double was OK ✓

(D) Adjust horizontal hair to optical axis of scope
Direct - Near Rod 5.214 Far Rod 5.30
Plunged - " 5.214 " 5.34
Adjusted with capstan to read Far Rod 5.32
Repeated as check ✓ OK

(E) Adjust standards
Plumb center of bolt top of wall to floor
direct & reverse - error 0.36.' Split error
elevated scope & adjusted with right
standard trunnion block capstan
Repeated until inst. checked ✓ OK

(F) Peg adjustment to telescope level vial
Setup exactly 1/2 way btwn A & B
A = 5.215   B = 5.316
Set at A
A = 5.20   B = 5.311 ≠ 5.301
adjusted scope bubble to read 5.30 at B

(G) Adjust vert circle
Level plate & scope bubble
Vertical circle reads 0°01'30"
Adjusted to read 0°00'00" ✓

Note:
Repeated all adjustments twice on
last trial All test OK without further
adjustment.

FIGURE 32

TEST & ADJUST TRANSIT

See Section 5-4

STANDARDIZATION STEEL TAPE #24657

Sept. 22, 1960

Weather - Warm & No Wind

H.C. - A.B. Smith

R.C. - D. Jones

1) Compared with base line in alley behind
office 1908 Beverly Blvd. Los Angeles, Calif.

2) Base line established as 100.000' @ 68° temp &
checked today with lower tape #16522 and
found to be correct - 100.000 @ 68°

Thermometers #8 & #12 in good agreement

Spring Balance #4 in good adjustment

Coefficient of expansion of steel tape
= 0.0000065'/1°/long.

Lufkin 100' steel tape #24657 - 1/4" wide, good-
-coated every ft. 0 to 100' with Add ft at
0 end, graduated in hundredths. - Tape
is in good condition

I. Tape #24657 supported throughout

TAPE READING	TEMP	TEMP. CORRECTION	LBS PULL	CORR. TAPE READING
99.985	75°	.0046	10#	99.990
99.984	75°	.0046	10#	99.989
99.986	75°	.0046	10#	99.991
99.985	75°	.0046	10#	99.990
			Σ	399.960 = 99.990
				4

Therefore, tape is 100.010' @ 68° with 10# & 11 fully
supported - or tape is 0.01' long

II. Tape #24657 supported at 0 & 100 only.

TAPE READING	TEMP	TEMP. CORRECTION	LBS PULL	CORR. TAPE READING
99.985	75°	.0046	24#	99.990
99.985	75°	.0046	24#	99.991
99.985	75°	.0046	24#	99.989
99.985	75°	.0046	24#	99.990
			Σ	399.960 = 99.990
				4

Tape is 100.010' @68° with 24# Pull supported
at 0 & 100'

FIGURE 33

STANDARDIZE TAPE

See Section 5-4

41

STANDARDIZATION SPRING BALANCE #12

Mean Correction	Balance #12	1st Test Balance # 17653	Correction	2nd Test Balance # 17653	
	0	0		0	
−1	5#	4	−1#	4	−1
0	10#	10	0	10	0
+1/4	15#	15 1/2	+1/2#	15	0
+1	20#	21	+1#	21	+1
+3 1/4	25#	28	+3#	28 1/2	+3 1/2
+4 3/4	30#	34 1/2	+4 1/2#	35	+5
+4	35#	39	+4#	39	+4
+4 3/4	40#	44 1/2	+4 1/2#	45	+5
+3 3/4	45#	49	+4#	48 1/2	+3 1/2
+4	50#	54	+4#	54	+4
+3 1/4	55#	58 1/2	+3 1/2#	58	+3
+4	60#	64	+4#	64	+4

Sept 22, 1960

Compared by
– A.B. Smith

PROCEDURE

Spring balance #12 was hooked to Ajax Dial Face Spring Balance #17653 which was locked in a vise & Pull was applied in 5# increments as per #12 scale. Both balances were read

Test done in field supply room, temp. 70°

Ajax balance #17653 was tested Feb 10, 1960 by L.A. City Bu. of weights & meas. and found to be correct to ± 1/16 lb. (See certificate dated Feb 10, 1960)

Balance #17653 graduated in 1/8 lb increments
Balance #12 graduated in 1 lb increments

Balance #17653 Read by – D. Jones
Balance #12 Read by – A.B. Smith.

FIGURE 34

STANDARDIZE SPRING BALANCE

See Section 5–4

55

DETERMINATION OF STADIA FACTOR K
BERGER TRANSIT #5328

Weather - Cold & Clear   Oct. 15, 1960

P.C. - A. Smith
T - B. Jones
H.C. - C. White
R.C. - D. Doe

Inst. at O

Point #	S	D	K
1	1.11	110.84	99.86
2	1.76	175.36	99.64
3	2.26	225.75	99.89
4	3.01	300.91	99.97
5	3.51	350.22	99.78
6	4.45	444.73	99.94
7	5.86	585.41	99.90
8	7.21	720.12	99.88
9	8.51	850.41	99.93
10	10.02	1001.40	99.94
		E =	998.73
		Mean K =	99.87

Note:

S = Rod intercept
D = Meas. distance
K = Stadia factor

$$K = \frac{D}{S}$$

Corrected stadia distance = .3 x K

K&E 13' Philadelphia Rod #A-16
Compared with K&E 100' tape - Std @ 68° & 24#
All points on line and on straight grade of less than 1%

Note: On stadia reading over 500' two targets were used.

FIGURE 35

DETERMINATION OF STADIA FACTOR
See Section 5-4

CALIBRATE BENDIX ECHO SOUNDER
No. B-378934-L.M.

Weather - Cold & Windy
Sea - 3' swells

Nov. 15 1959
P.C. - A. Smith
CH - B. Jones

L = Lead line depth
C = Depth of transducer head below W.L.
S = Depth read from Echo Sounder graph
S+C = Depth of water per Echo Sounder + C
K = Constant to apply to chart value

Sounder mounted in 28' Motor Launch "MARY-J" Transducer mounted on starboard side, midship - head is 2.1' below water line.

Notes.
1. Test made behind breakwater in calm water.
2. Steel lead line with 50# paravane
3. Hard sand bottom
4. No correction made for water temp. or salinity

Checks OK with morning readings

Test 9:00 PM before starting job

S	C	S+C	L	K
10.5	+2.1	12.6	12.9	+0.3
18.2	+2.1	20.3	20.8	+0.5
24.3	+2.1	26.4	27.0	+0.6
30.0	+2.1	32.1	32.8	+0.7
38.7	+2.1	40.8	41.5	+0.7
49.4	+2.1	51.5	52.0	+0.5
56.2	+2.1	58.3	58.9	+0.6

$$\text{Mean } K = \frac{3.9}{7} = +0.6$$

Apply +0.6 to graph readings S+C

Test 4:45 P.M. after completing job

S	C	S+C	L	K
11.4	+2.1	13.5	14.1	+0.6
19.1	+2.1	21.2	21.6	+0.4
23.6	+2.1	25.7	26.0	+0.3
33.3	+2.1	35.4	35.8	+0.4
41.5	+2.1	43.6	44.3	+0.7
50.7	+2.1	52.8	53.2	+0.4
57.3	+2.1	59.4	60.1	+0.7

$$\text{Mean } K = \frac{3.5}{7} = +0.5$$

FIGURE 36
CALIBRATE ECHO SOUNDER
See Section 5-4

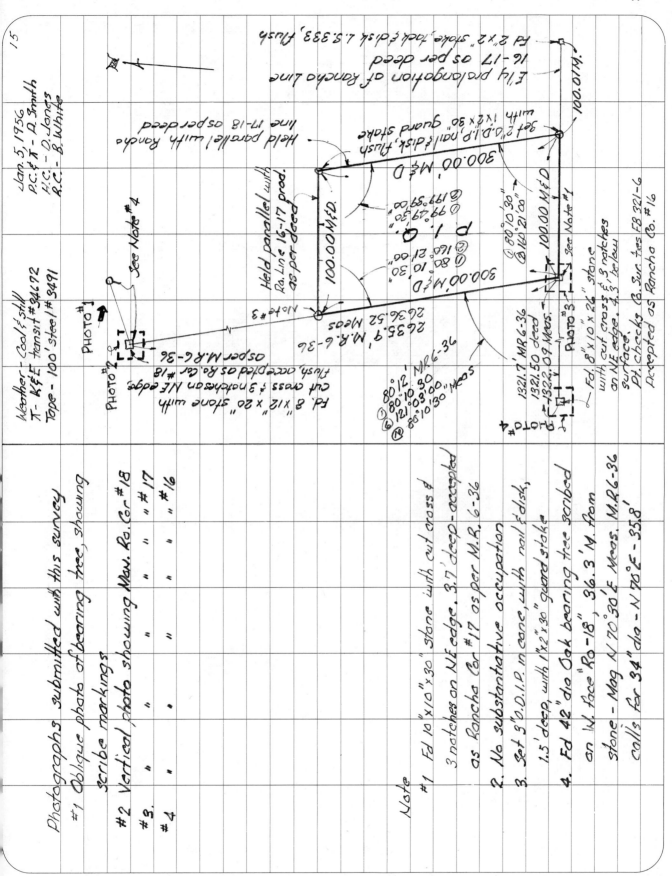

FIGURE 37

PHOTOGRAPHY AS SUPPORT DATA
See Section 5-5

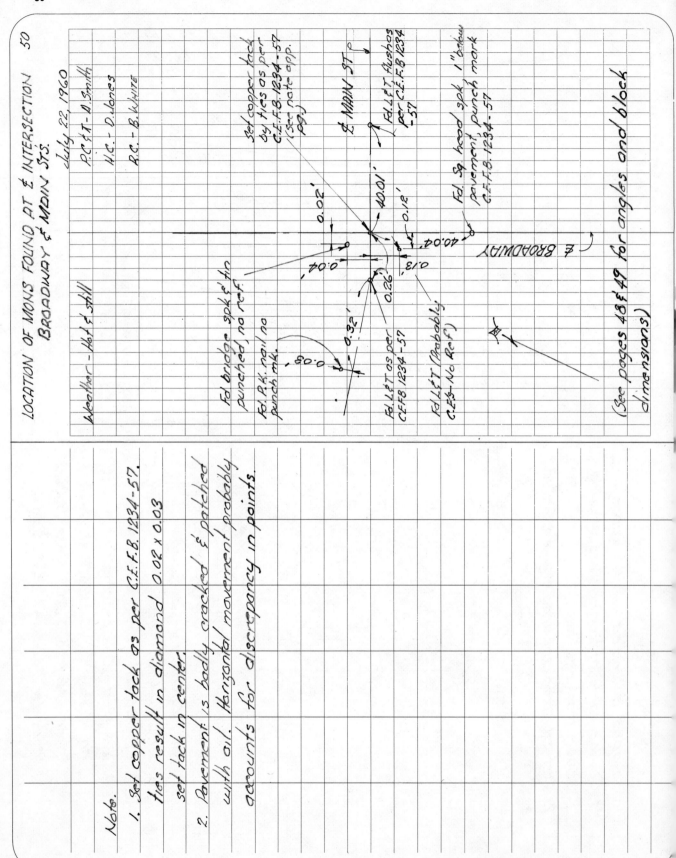

80

FIGURE 38

TAPING NOTES, SHORT DISTANCES
LOCATION OF FOUND MONUMENTS
See Section 6-2

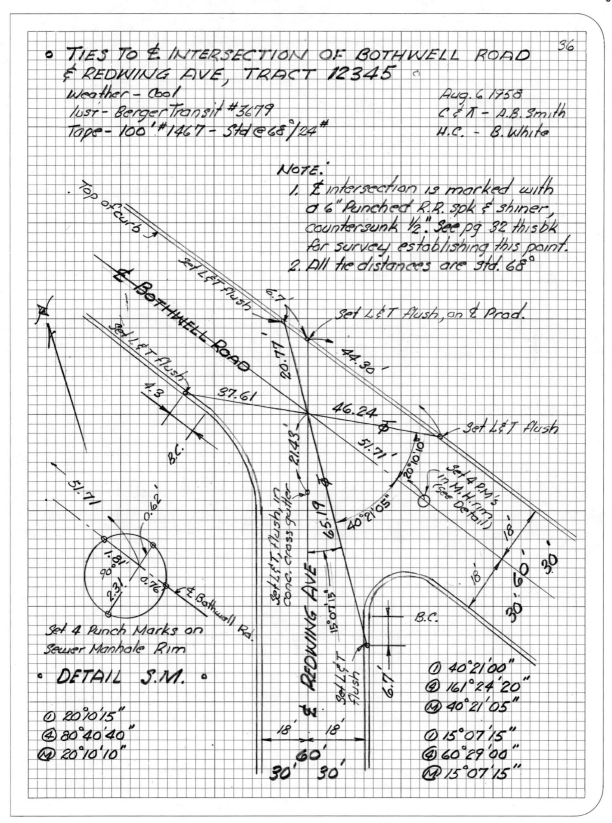

FIGURE 39

TAPING NOTES, SHORT DISTANCES
CENTERLINE TIES
See Section 6-2

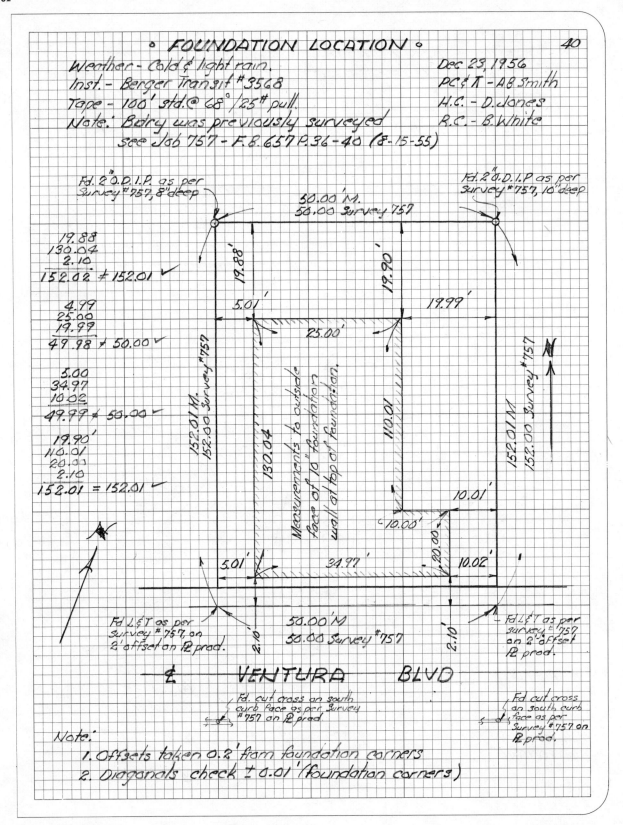

FIGURE 40

TAPING NOTES, SHORT DISTANCES
FOUNDATION LOCATION
See Section 6-2

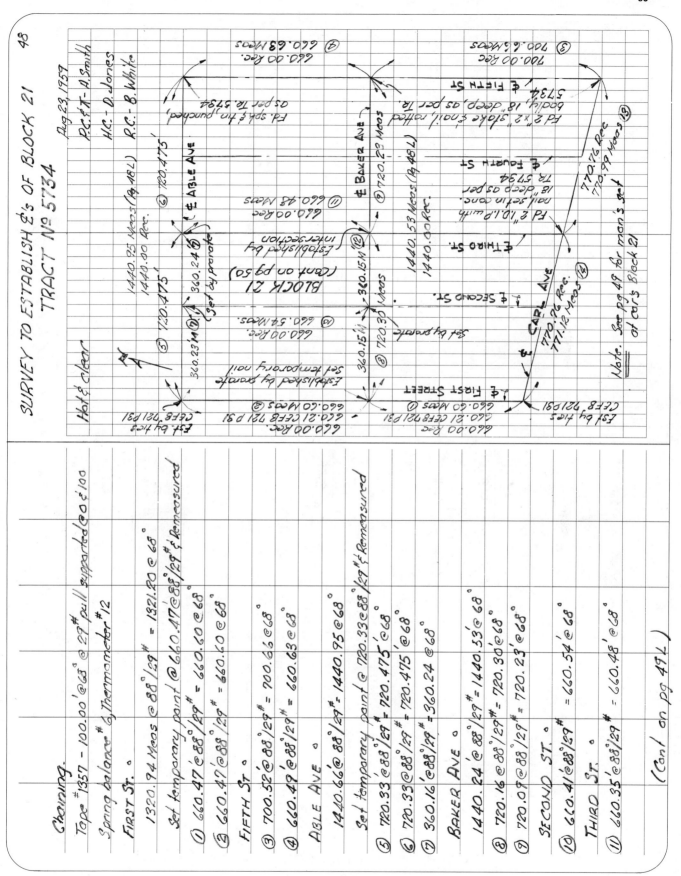

FIGURE 41

TAPING NOTES, LONG DISTANCES
CENTERLINE SURVEY
See Section 6-2

84

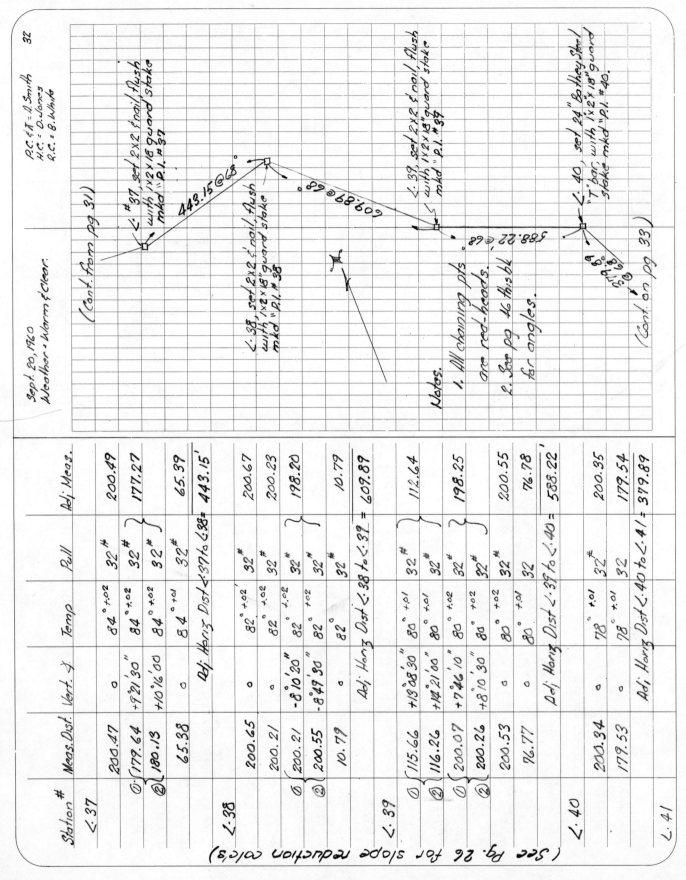

FIGURE 42

TAPING NOTES, LONG DISTANCES
TRAVERSE CHAINING
See Section 6-2

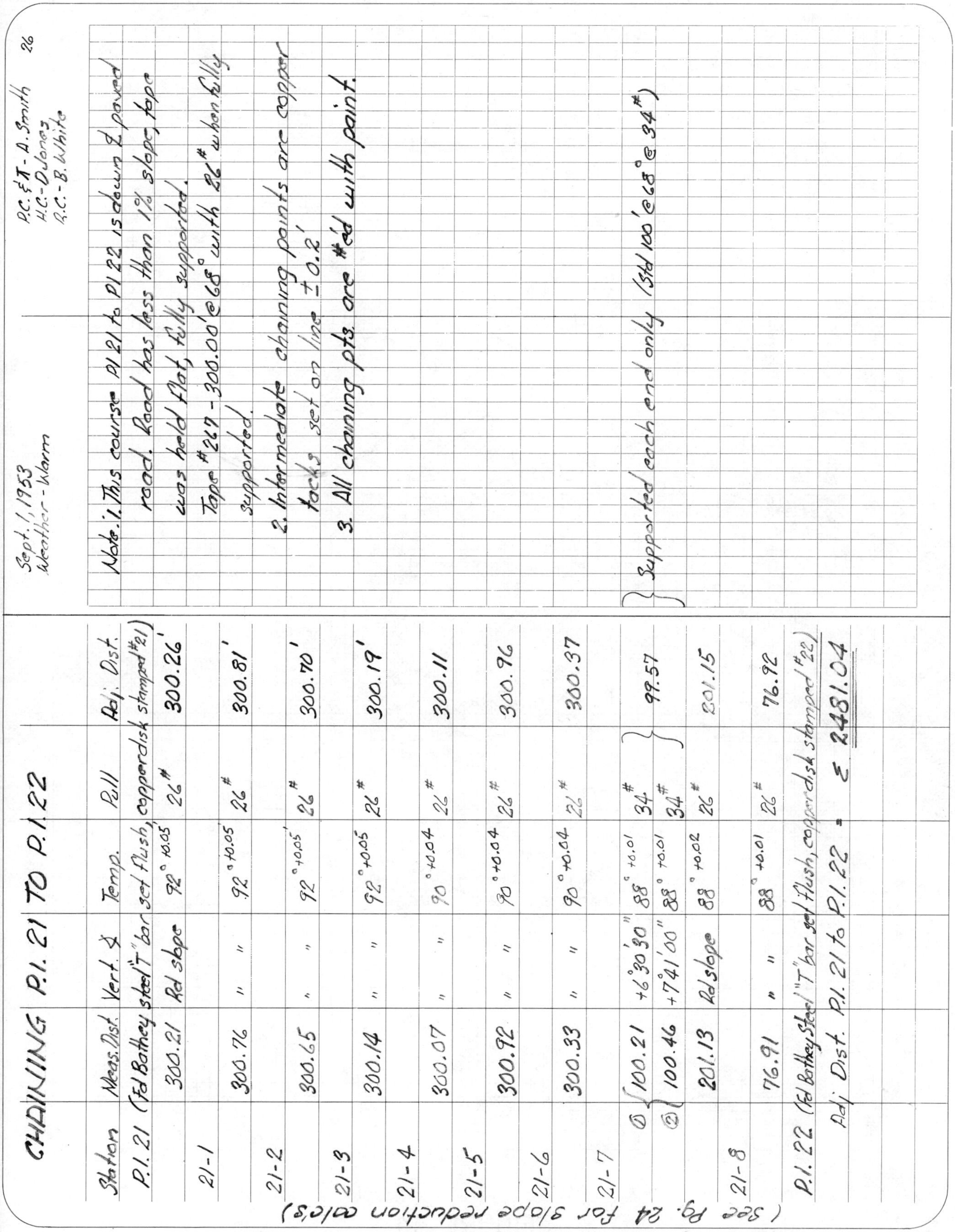

26

Sept. 1, 1953
Weather - Warm

P.C. ¢ T. - A. Smith
H.C. - D. Jones
R.C. - B. White

CHAINING P.I. 21 TO P.I. 22

Station	Meas. Dist.	Vert. ∡	Temp.	Pull	Adj. Dist.
P.I. 21 (Fd Bothey steel "T" bar set flush, copper disk stamped #21)					
	300.21	Rd slope	$92°+0.05$	26#	300.26'
21-1	300.76	"	$92°+0.05$	26#	300.81'
21-2	300.65	"	$92°+0.05$	26#	300.70'
21-3	300.14	"	$92°+0.005$	26#	300.19'
21-4	300.07	"	$90°+0.04$	26#	300.11'
21-5	300.92	"	$90°+0.04$	26#	300.96'
21-6	300.33	"	$90°+0.04$	26#	300.37'
21-7	① 100.21	$+6°30'30"$	$88°+0.01$	34#	97.57
	② 100.46	$+7°41'00"$	$88°+0.01$	34#	
21-8	201.13	Rd slope	$88°+0.02$	26#	201.15
	76.91	"	$88°+0.01$	26#	76.92
P.I. 22 (Fd Bothey Steel "T" bar set flush, copper disk stamped #22)					
Adj. Dist. P.I. 21 to P.I. 22 =					Σ 2481.04

(See Pg. 24 for slope reduction calc's)

Note: 1. This course P.I. 21 to P.I. 22 is down ¢ paved road. Road has less than 1% slope. Tape was held flat, fully supported. Tape #217 - 300.00' @ 68° with 26# when fully supported.

2. Intermediate chaining points are copper tacks set on line ±0.2'

3. All chaining pts are #'d with paint.

{ Supported each end only (Std 100' @ 68° @ 34#)

FIGURE 43

TAPING NOTES, LONG DISTANCES
TRAVERSE CHAINING
See Section 6-2

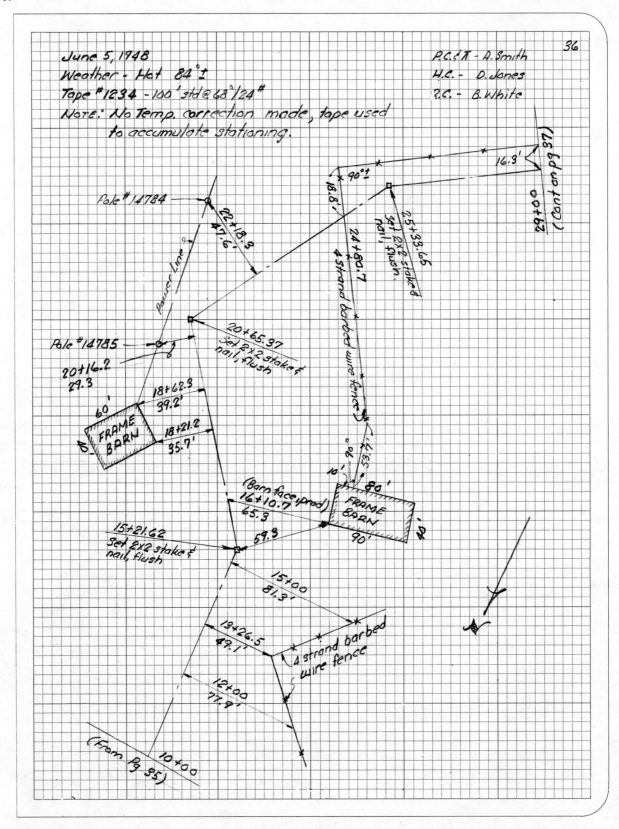

June 5, 1948
Weather - Hot 84°±
Tape #1234 - 100' std @ 68°/24#
Note: No Temp. correction made, tape used
      to accumulate stationing.

P.C.¢π - A.Smith
H.C. - D.Jones
?.C. - B.White

Pole #14784

22+18.3
47.6'

Power Line

Pole #14785

20+16.2
29.3

60'

18+62.3
39.2'

FRAME
BARN

18+21.2
35.7'

20+65.37
Set 2x2 stake ¢
nail, flush

90°±

18.8'

24+60.7
4 strand barbed wire fence

25+33.65
Set 2x2 stake ¢
nail, flush

16.3'

29+00
(Cont on pg 37)

90°

53.7'

10'

80'

90'

FRAME
BARN

(Barn face prod)
16+10.7
65.3

59.3

15+21.62
Set 2x2 stake ¢
nail, flush

15+00
81.3'

13+26.5
49.1'

4 strand barbed
wire fence

N

12+00
77.9'

(From pg 35)
10+00

FIGURE 44

TAPING NOTES, STATIONING
TRAVERSE CHAINING
See Section 6-2

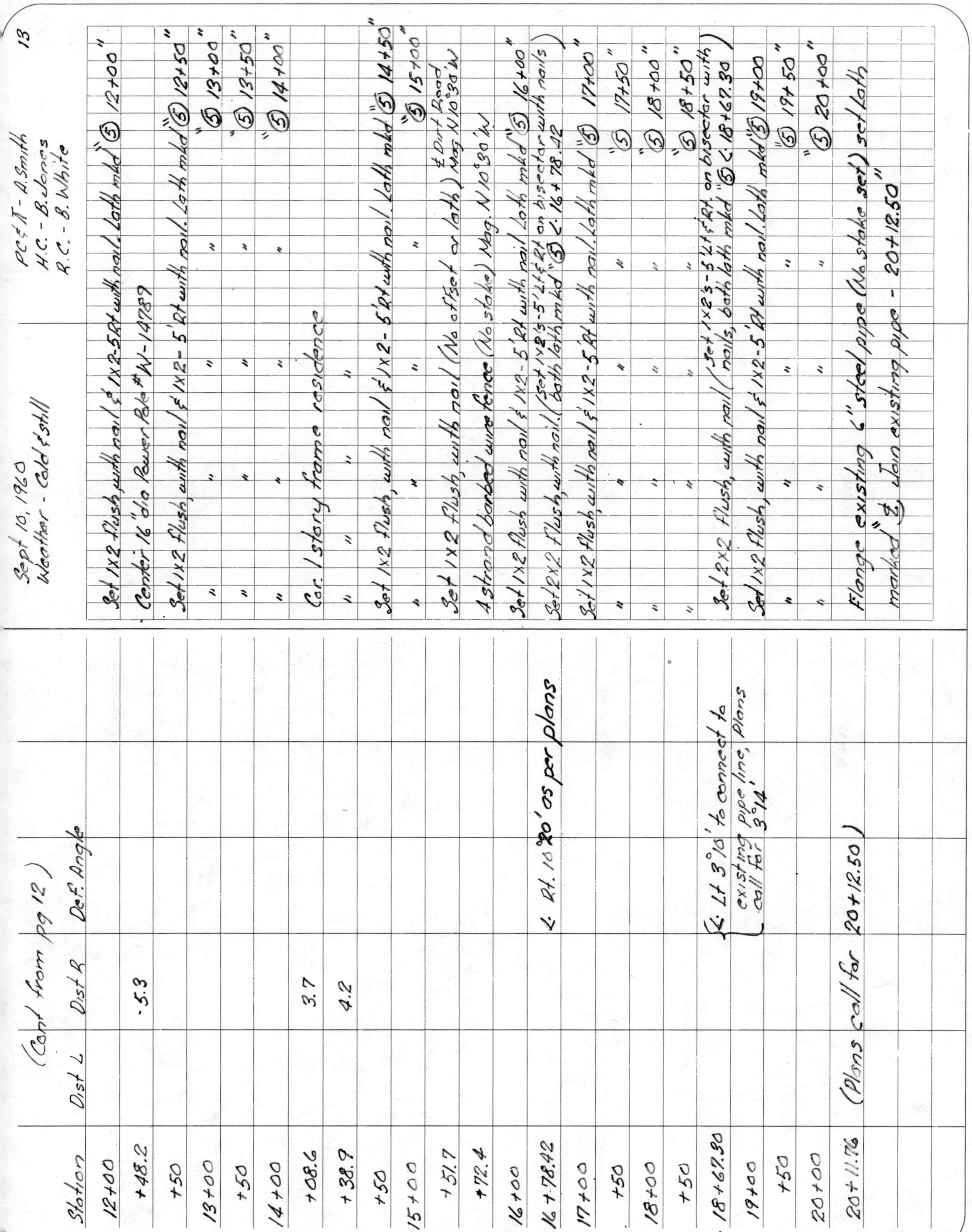

Left page — Field notes

PC & IT - A. Smith
H.C. - B. Jones
R.C. - B. White

Sept 10, 1960
Weather - Cold & still

Description	Station
Set 1x2 flush with nail & 1x2-5 Rt with nail. Lath mkd "	⑤ 12+00 "
Center 16" dia Power Pole # W-14789	
Set 1x2 flush, with nail & 1x2 - 5' Rt with nail. Lath "	⑥ 12+50 "
"	② 13+00 "
"	⑤ 13+50 "
"	⑤ 14+00 "
Cor. 1 story frame residence	
Set 1x2 flush, with nail & 1x2 - 5 Rt with nail. Lath mkd "	⑤ 14+50 "
"	⑤ 15+00 "
Set 1x2 flush, with nail (No offset or lath) ℄ Dirt Road Mag. N 10°30' W	
A strand barbed wire fence (No stake)	
Set 1x2 flush with nail & 1x2-5 Rt with nail. Lath mkd "	⑤ 16+00 "
Set 2x2 flush, with nail. (Set 1x2 3-5'Rt & 24 Rt on bisector with nails) "	⑤ ℄ 16+78.12
Set 1x2 flush, with nail & 1x2-5 Rt with nail. Lath mkd "	⑤ 17+00 "
"	⑤ 17+50 "
"	⑥ 18+00 "
"	⑤ 18+50 "
Set 2x2 flush, with nail. (Set 1x2-3-5'2.4'Rt on bisector with nails, both lath mkd "	⑤ ℄ 18+67.30)
Set 1x2 flush, with nail & 1x2-5 Rt with nail. Lath mkd "	⑤ 19+00 "
"	⑤ 19+50 "
"	⑤ 20+00 "
Flange existing 6" steel pipe (No stake set) set lath marked ℄ Join existing pipe - 20+12.50	

Right page — Stationing table

(Cont from pg 12)

Station	Dist L	Dist R	Def. Angle
12+00			
+48.2		-5.3	
+50			
13+00			
+50			
14+00			
+08.6		3.7	
+38.9		4.2	
+50			
15+00			
+51.7			
+72.4			
16+00			
℄ 16+78.42			∠ Rt. 10°20' 05 per plans
17+00			
+50			
18+00			
+50			
℄ 18+67.30			∠ Lt 3°10' to connect to existing pipe line, Plans call for 3°14'
19+00			
+50			
20+00			
20+11.76			(Plans call for 20+12.50)

FIGURE 45

TAPING NOTES, STATIONING
PIPELINE LAYOUT
See Section 6-2

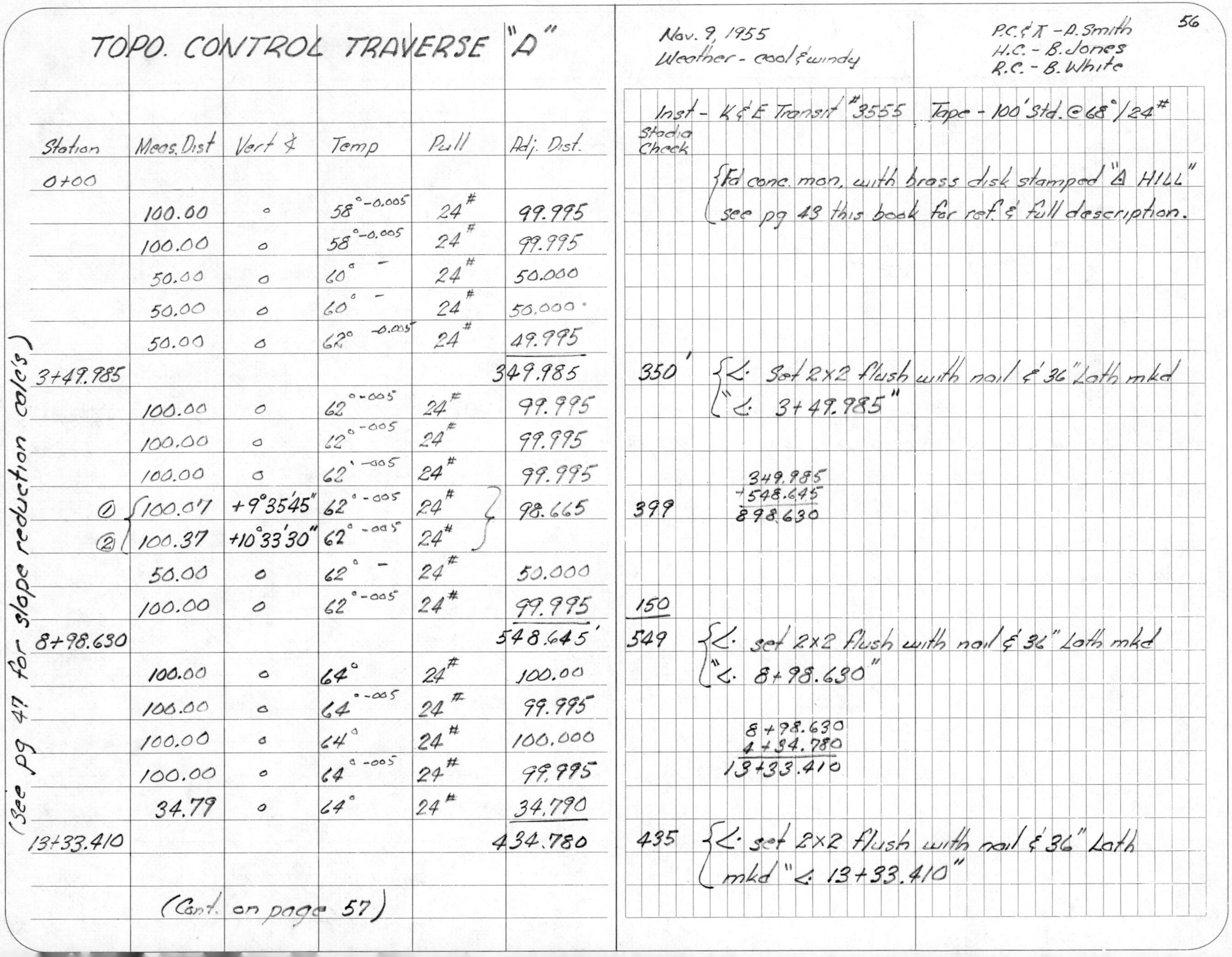

56

TOPO. CONTROL TRAVERSE "D"

Nov. 9, 1955
Weather - cool & windy

Inst - K&E Transit #3555   Tape - 100' Std. @ 68°/24#

P.C. & ∆ - D. Smith
H.C. - B. Jones
R.C. - B. White

Station	Meas. Dist.	Vert ∢	Temp	Pull	Adj. Dist.
0+00					
	100.00	0	58° -0.005	24#	99.995
	100.00	0	58° -0.005	24#	99.995
	50.00	0	60° -	24#	50.000
	50.00	0	60° -	24#	50.000
	50.00	0	62° -0.005	24#	49.995
3+49.985					349.985
	100.00	0	62° -0.005	24#	99.995
	100.00	0	62° -0.005	24#	99.995
	100.00	0	62° -0.005	24#	99.995
	① 100.01	+9°35'45"	62° -0.005	24#	98.665
	② 100.37	+10°33'30"	62° -0.005	24#	
	50.00	0	62° -	24#	50.000
	100.00	0	62° -0.005	24#	99.995
8+98.630					548.645
	100.00	0	64° -0.005	24#	100.00
	100.00	0	64° -0.005	24#	99.995
	100.00	0	64° -0.005	24#	100.000
	100.00	0	64° -0.005	24#	99.995
	34.79	0	64° -	24#	34.790
13+33.410					434.780

Stadia Check

{Fd conc. mon. with brass disk stamped △ HILL "see pg 43 this book for ref & full description.

350' {℄ Set 2x2 flush with nail & 36" lath mkd "℄ 3+49.985"

349.985
-548.645
898.630

399

150

549 {℄ set 2x2 flush with nail & 36" lath mkd "℄ 8+98.630"

8+98.630
4+34.780
13+33.410

435 {℄ set 2x2 flush with nail & 36" lath mkd "℄ 13+33.410"

(Cont. on page 57)

(See pg 47 for slope reduction calc's)

FIGURE 46

TAPING NOTES, STATIONING
TRAVERSE CHAINING
See Section 6-2

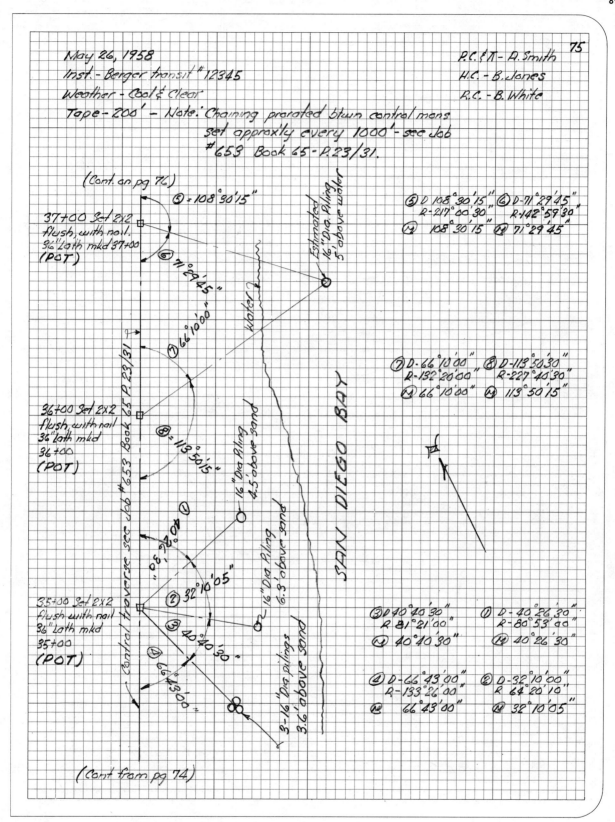

FIGURE 47

ANGLE NOTES, RANDOM ANGLES
LOCATION BY INTERSECTION
See Section 6-3

90

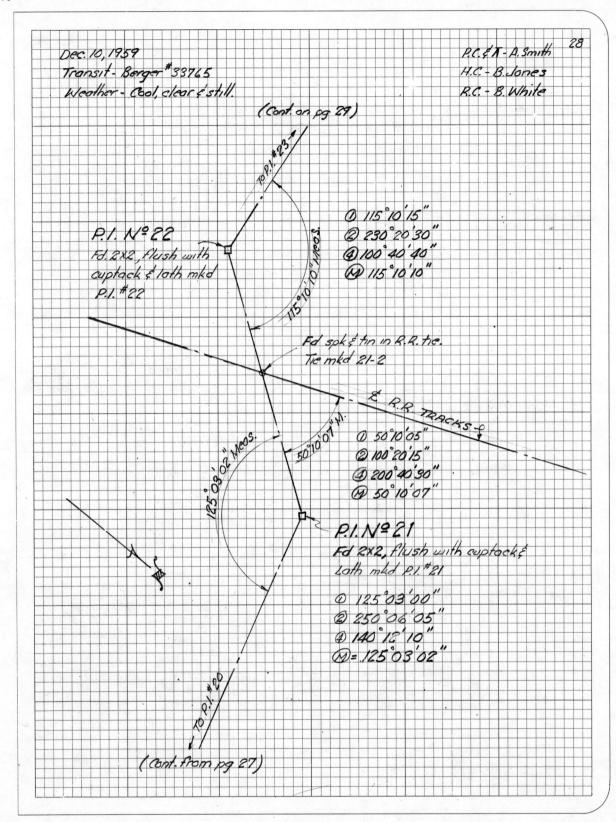

FIGURE 48

ANGLE NOTES, TRAVERSE

See Section 6-3

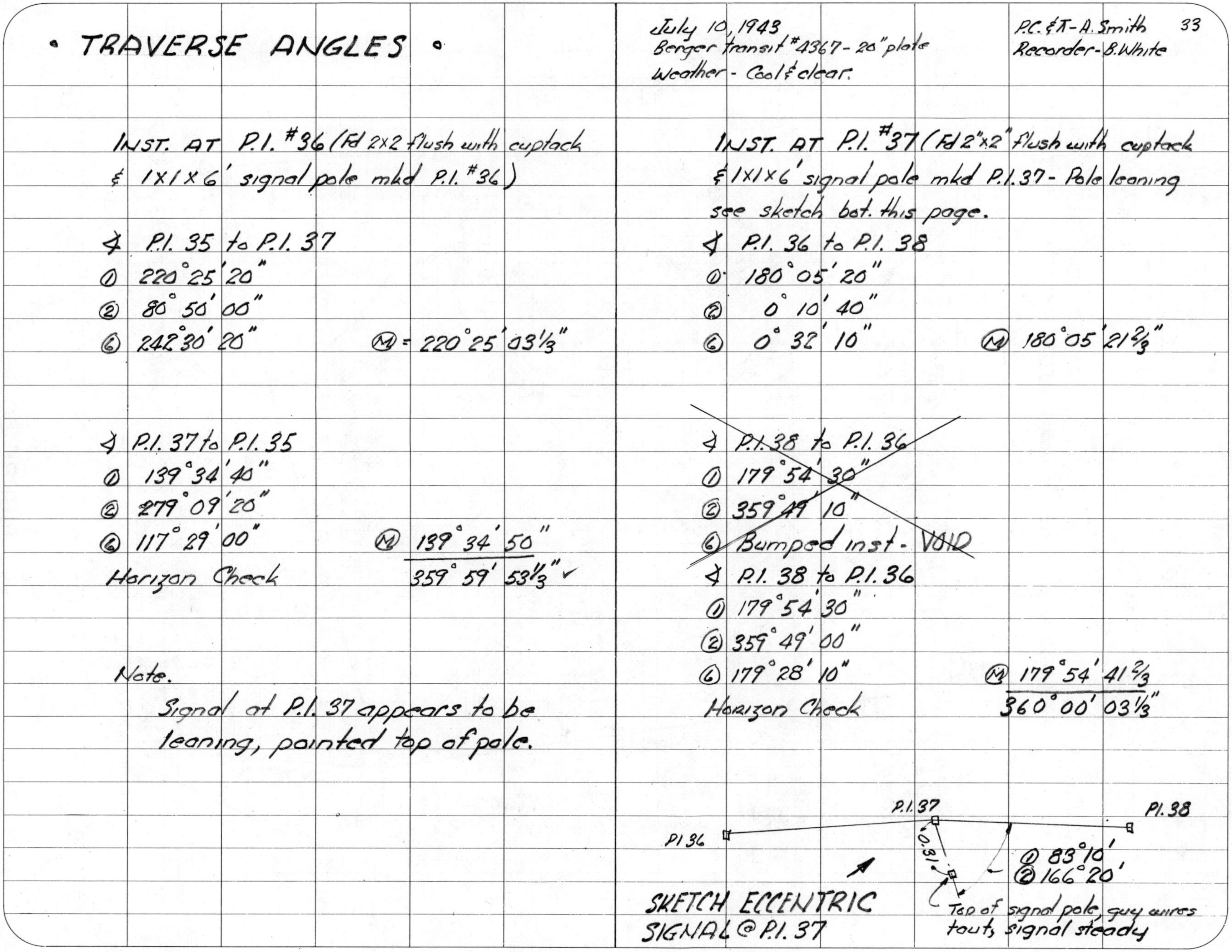

33

• TRAVERSE ANGLES °

	July 10, 1943	Plc. & A. Smith
Berger transit #4367 - 20" plate	Recorder - B. White
Weather - Cool & clear.	

INST. AT P.I. #36 (Fd 2"x2" flush with cuplack)
℄ 1 x 1 x 6' signal pole mkd P.I. #36.)

∠ P.I. 35 to P.I. 37
① 220° 25' 20"
② 80° 50' 00"
Ⓒ 242° 30' 20"
Ⓜ = 220° 25' 03⅓"

∠ P.I. 37 to P.I. 35
① 139° 34' 40"
② 279° 09' 20"
Ⓒ 117° 29' 00"
Horizon Check

Ⓜ 139° 34' 50"
    359° 59' 53⅓" ✓

Note.
Signal at P.I. 37 appears to be
leaning, painted top of pole.

INST. AT P.I. #37 (Fd 2"x2" flush with cuplack)
℄ 1 x 1 x 6' signal pole mkd P.I. 37 - Pole leaning
see sketch bot. this page.

∠ P.I. 36 to P.I. 38
① 180° 05' 20"
② 0° 10' 40"
Ⓒ 0° 32' 10"
Ⓜ 180° 05' 21⅔"

∠ P.I. 38 to P.I. 36
① 179° 54' 30"
② 359° 49' 10"
Ⓒ Bumped inst. - VOID

∠ P.I. 38 to P.I. 36
① 179° 54' 30"
② 357° 49' 00"
Ⓒ 179° 28' 10"
Horizon Check

Ⓜ 179° 54' 41⅔"
    360° 00' 03⅓"

SKETCH ECCENTRIC
SIGNAL @ P.I. 37

P.I. 38
① 83° 10'
② 166° 20'
Top of signal pole guy wires
Top of signal pole not steady
P.I. 37   0.31'
P.I. 36

FIGURE 49

ANGLE NOTES, TRAVERSE
See Section 6-3

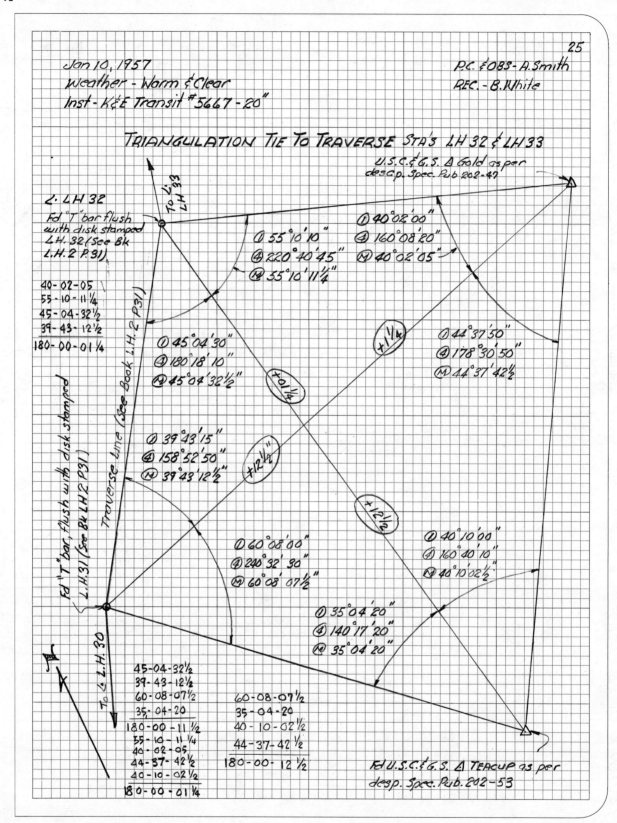

FIGURE 50

ANGLE NOTES, TRIANGULATION
See Section 6−3

OCCUPY STATION   S.R. #88 - TRIANGULATION
(Fd point in good condition as described in Bk S.R.3 P.26)

Inst - Wild T-2 #12479
Weather - Cool, clear, high overcast

P.C. & Obs. - A. Smith 32
Recorder - B. White

POSITION	OBJECT OBSERVED	OBSERVATION	MEAN D & R		ABSTRACT OF DIRECTIONS ①	②	③	④	MEAN
POSITION 1	S.R.86	D 0-00-15	00-00-18	0°00'	00"	00"	00"	00"(circled)	00"
		R 180°00 21							
	S.R.85	D 50-04-31	50-04-35	50°04'	17"	15"	20"	18"	18"
		R 230-04-39							
Reobserved	S.R.87	D 89-09-40	89-09-42	89°09'	24"(07)	05"	10"	09"	08"
		R 269-09-44							
	S.R.89	D 134-07-16	134-07-18	134°07'	00"	56̄"	02"	02"	00"
		R 214-07-21							
	S.R.90	D 181-17-21	181-17-22	181°17'	04"	(57")7 0̄0 3̄58 06"	06"	09"	05"(04)
		R 01-17-23							
POSITION 2	S.R.86	D 45-02-30	45-02-34	0°00'	00"				
		R 220-02-37							
	S.R.85	D 095-06-47	95-06-49	50°04	15"				
		R 275-06-51							
	S.R.87	D 134-11-38	134-11-39	89°09'	05"				
		R 314-11-40							
	S.R.89	D 0179-09-30	179-09-30	134°06'	56"				
		R 359-09-30							
	S.R.90	D 226-19-34	226-19-34	181°17'	00"				
		R 46-19-35							

(Cont on pg 33)

FIGURE 51

ANGLE NOTES, TRIANGULATION
See Section 6-3

94

POSITION	OBJECT OBSERVED	OBSERVATION	MEAN D & R	DIRECTIONS
POSITION 3	S.R.86	D 90-05-00 / R 270-05-01	90-05-01	0° 00' 00"
	S.R.85	D 140-09-17 / R 320-09-25	140-09-21	50° 04' 20"
	S.R.87	D 179-14-08 / R 359-14-13	179-14-11	89° 09' 10"
	S.R.89	D 224-12-00 / R 44-12-06	224-12-03	134° 07' 02"
	S.R.90	D 271-22-05 / R 91-22-10	271-22-07	181° 17' 06"
POSITION 4	S.R.86	D 135-07-34 / R 315-07-38	135-07-36	0° 00' 00"
	S.R.85	D 185-11-51 / R 05-11-56	185-11-54	50° 04' 18"
	S.R.87	D 224-16-42 / R 44-16-48	224-16-45	89° 09' 09"
	S.R.89	D 269-14-36 / R 89-14-40	269-14-38	134° 07' 02"
	S.R.90	D 316-24-41 / R 136-24-49	316-24-45	181° 17' 09"

(Cont. on pg 34)

FIGURE 52

ANGLE NOTES, TRIANGULATION
See Section 6—3

34

POSITION	OBJECT OBSERVED	OBSERVATION	MEAN D&R	DIRECTION
POSITION 1 REOBSERVED				
	S.R.86	D 00-00-05	00-00-07	
		R 180-00-08		
	S.R.87	D 89-09-13	89-09-14	89°09'07"
		R 269-09-15		
POSITION 2 REOBSERVED				
	S.R.86	D 45-02-35	45-02-36	
		R 225-02-37		
	S.R.90	D 226-19-33	226-19-33	181°16'57"
		R 46-19-33		

VERTICAL ANGLES   OCCUPY STATION S.R. #88   HI = 4.93

STATION	SIGNAL HEIGHT	1ST SET Direct	Reverse	DIRECT ADJ	2ND SET Direct	REVERSE	DIRECT ADJ	DIRECT MEAN	VERT ∢
S.R.86	10.12'	91-15-24	268-44-48	91-15-18	15-30	44-36	15-27	91-15-23	-1°15'23"
S.R.85	16.00'	89-20-22	270-35-44	89-20-19	20-20	39-42	20-19	89-20-19	+0°39'41"
S.R.87	10.05'	76-18-33	283-41-33	76-18-30	18-34	41-32	18-31	76-18-31	+13°41'29"
S.R.89	10.06'	81-43-12	278-16-52	81-43-10	43-08	16-56	43-06	81-43-08	+8°16'52"
S.R.90	16.03'	90-53-03	269-06-59	90-53-02	53-05	06-57	53-04	90-53-03	-0°53'03"

FIGURE 53

ANGLE NOTES, TRIANGULATION

See Section 6-3

28

DOUBLE ROD LEVEL CIRCUIT – EST. T.B.M.'s ALONG CO. RD. NO. 15 · MODESTO TO 14 MI. WEST

° A ROD °

Sight	B.S.	H.I.	F.S.	Elev.	Field Adj. Elevation
B.M. #332 = Cut cross in top of conc. footing N.E. Tower leg of A.B.C. Co's 50,000 Gal. storage tank (See Level Bk 2 P. 31-66 699 for full description)	5.21	531.52			526.31
T.P. A-1	4.99	531.90	4.61	526.91	
T.P. A-2	4.72	531.11	5.51	526.39	
T.P. A-3	4.31	530.51	4.91	526.20	
T.P. A-4	12.61	540.00	3.12	527.39	
T.B.M. #1 Set R.R. spk & disk in top of log post of guard rail, most N'ly post on west side of road. County Road #15 at Branch Creek Bridge			4.13	535.87	535.86
T.P. A-5	4.72	540.58	4.48	536.10	
T.P. A-6	4.71	540.81	9.32	531.49	

(Cont. on pg 29)

Berger Dumpy Level – 3567 Weather - Cool & clear

Jan. 3, 1959 P.C.& T. – P. Smith A.φ – B. Jones B.φ – B. White

° B ROD °

Sight	B.S.	H.I.	F.S.	Elev.	Field Adj. Elevation
B.M. #332					526.31
T.P. B-1	5.21	531.52	5.22	526.30	
T.P. B-2	5.60	531.90	6.01 / 5.64	525.89	
T.P. B-3	5.21	531.10	5.82	525.28	
T.P. B-4	5.21	530.49	4.23	526.26	
T.B.M. #1	13.72	539.98	4.13	535.85	535.86
T.P. B-5	4.72	540.58	5.21	535.37	
T.P. B-6	5.43	540.80	10.60	530.20	

(Cont. on pg 29)

FIGURE 54

LEVEL NOTES
SINGLE WIRE-DOUBLE RODDED CIRCUIT
See Section 6-4

14

FLY LEVELS PHOTO. CONTROL
FISH CANYON RD - FOOTHILL TO 8 Mi N.

FEET

Sight	B.S.	H.I.	F.S.	Elev.		Sight	B.S.	H.I.	F.S.	Elev.	Check
B.M. 108NJ	Std. U.S.G.S. B.M. Mon as			691.32		B.N. 108NJ				230.44 Yds.	
	per U.S.G.S. Description form										
	9-1078-A Stamped "10 BNH-1949"										
	-691"										
	9.96	701.28					3.32	233.76			
P#1			1.01	700.27					0.34	233.42	
	14.36	714.63					4.78	238.20			
P#2			1.31	713.32					0.44	237.76	
	15.99	729.31					5.33	243.09			
P#3			1.21	728.10					0.40	242.69	
	15.20	743.30					5.07	247.76			
P.P.#4-17B			2.13	741.17					0.74	249.02 ×3 = 741.09	
P.P.#4-17B - 4' Dia white painted circle & Road. See Photo 4-17 for identification & description											
	14.26	755.43					4.75	251.77			
P#4			1.67	753.76					0.56	251.21	
	15.03	768.79					5.01	256.22			
P#5			1.30	767.49					0.43	255.79	
	15.61	783.10					5.20	260.99			
P#6			1.08	782.02					0.36	260.63	
	15.50	797.52					5.17	265.80			
			1.32	796.20					0.44	265.36 ×3 = 796.08	

(Cont. on pg 15)   (Cont. on pg 15)

YARD CHECK

Feb 14, 1951
P.C. & T - D. Smith
Rec - B. Jones
Rod - B. White

Weather - Hot & hazy
Inst - K & E Dumpy #7373
Rod - Single Piece 16'
F.t on front Yds on back

FIGURE 55

LEVEL NOTES
SINGLE WIRE-DOUBLE FACE ROD
See Section 6-4

98

87

Aug. 14 1958 - 8:15 A.M.
Rodⁿ - A. Smith
Inst. Berger #3777
φ - B. White

11:20 AM

(Cont. from pg. 86)

T.B.M. L.A.F. 21 TO L.A.F. 22 — FORWARD RUN

	+	−
T.B.M.LAF 21 (Desc Pg 36)	5.127	
	5.723	5.378
	6.173	5.632
	5.372	5.882
	5.117	4.638
	6.831	3.124
	8.721	2.679
	10.333	2.793
	14.721	3.762
	11.936	3.292
	10.845	3.281
	11.755	2.118
	8.911	3.211
	4.213	4.786
	5.865	5.218
		5.345
T.B.M.LAF 22 (Desc Pg 37)	Σ +121.143	−61.439

Diff.Elev + 59.704 + 0.008 = 59.712 Adj. Diff.
− 57.721 − 0.009 = 59.712 "
0.017 Divergence

Elev. T.B.M. LAF 21 (From pg 86) = 1037.629
+ 59.712
Elev. T.B.M. LAF 22 = 1097.341
(Cont. on Pg. 88)

T.B.M. L.A.F. 22 TO L.A.F. 21 — BACKWARD RUN

	+	−
T.B.M.LAF 22 (Desc Pg 37)	5.453	
	5.281	5.635
	5.101	4.231
	3.112	9.003
	2.308	11.811
	3.013	11.763
	3.522	12.172
	4.121	15.072
	2.479	10.132
	2.779	8.812
	3.562	6.913
	4.468	5.011
	5.891	5.377
	5.636	5.982
	5.111	5.632
T.B.M.LAF 21 (Desc Pg 36)		4.012
	Σ + 61.837	−121.558

Diff.Elev + 61.837
− 57.721

Note: Shots balanced exactly with measuring wheel. Total loop distance = 11840 Ft.

FIGURE 56

LEVEL NOTES
SINGLE WIRE-FORWARD & BACKWARD RUNS
See Section 6-4

PRECISE LEVELS
FORWARD RUN - B.M. T-4 TO B.M. T-5

Inst. - Berger level #22508
Sun - South & bright, Wind-calm
(See Book 54-3 Pg.16 for B.M. descr.)

Oct 15, 1957   8:00 A.M.   16
Party
P.C.& Rec.-A.Smith
Inst.-B.Jones
① Rod-C.White   ② Rod-D.Fitz

STATION/ROD	THREADS BACKSIGHT	MEAN	THREAD INTERVAL	SUM OF INTERVALS	FT. CHECK	TEMP/ROD	THREADS FORESIGHT	MEAN	THREAD INTERVAL	SUM OF INTERVALS	FT. CHECK
1 ②	3609		89				0533		85		
	3520	3520.0	89			43 ①	0448	0448.0	85		
	3431.		178	178			0363		170	170	
2 ①	2670		101				2435		101		
	2569	2569.7	99			43° ②	2334	2334.3	100		
	2470		200	378			2234		201	371	
3 ②	2156		103				2232		102		
	2053	2053.0	103			43° ①	2130	2130.3	101		
	1950		206	584			2029		203	574	
4 ①	1968		119				2001		121		
	1849	1848.7	120			44° ②	1880	1880.0	121		
	1729		239	823			1759		242	816	
5 ②	1735		105				1976		104		
	1630	1629.7	106			44° ①	1872	1872.0	104		
	1524		211	1034			1768		208	1024	
6 ①	1843		33				1747		40		
	1810	1810.0	33			45° ②	1707	1707.0	40		
	1777		66	1100			1667		80	1104	
				1104				10371.6			

-13431.1
+10377.6
Forward Run = -3060.5
2.204 K/m.

(Cont on pg. 17)

FIGURE 57

LEVEL NOTES
PRECISE-THREE WIRE
See Section 6-4

PRECISE LEVELS °
BACKWARD RUN – B.M. T-5 TO B.M. T-4

Oct 15, 1957 8:30 A.M.   17

(Cont. from pg. 16)

Sta.	Backward Run readings	Mean	Intervals	Sum		Cont. readings	Mean	Intervals	Sum
7 ②	1917		110			2028		108	
	1807	1807.0	110		45° ①	1920	1919.7	109	
	1697		220	220		1811		217	217
8 ①	1990		111			1969		109	
	1879	1878.7	112		45° ②	1860	1860.0	109	
	1767		223	443		1751		218	435
9 ②	2222		102			2147		104	
	2120	2120.0	102		46° ①	2043	2043.3	103	
	2018		204	647		1940		207	642
10 ①	2335		101			2570		101	
	2234	2234.5	100		46° ②	2469	2469.7	99	
	2134		201	848		2370		200	842
11 ②	0682		98			3474		102	
	0584	0584.0	98		46° ①	3392	3392.0	102	
	0486		196	1044		3270		204	1046

8624.0

1046
2.090 Klm
2.204 Klm
Total length = 4.294 Klm

+ 11684.7
− 8624.0
+ 3060.7   Backward Run
− 3060.5   Forward Run
+ 0.2 Mm. Divergence

FIGURE 58

LEVEL NOTES
PRECISE – THREE WIRE
See Section 6–4

From B.M. 34 To B.M. 35   Weather - cool   June 28, 1959 10:00AM to 10:30AM

Station	Backsight Right Mic	Left	Stadia		Foresight Right Mic	Left	Stadia		Remarks
From Pg 17	1563 004		275		1003 976		270		
9	123 795	424.5	142/104 38		169 045	470.5	188/150 38		
10	130 629	431.5	149/111 38		173 192	474.5	192/154 38		
11	132 335	433.5	150/114 36		179 848	480.5	199/159 40		
12	141 100	442.5	160/122 38		198 457	499.5	221/183 38		
13	130 333	431.5	149/111 38		159 519	460.5	178/140 38		
14	137 598	438.5	156/118 38		171 370	472.5	190/152 38		
15	126 546	427.5	146/108 38		178 411	479.5	195/157 38		
16	118 853	419.5	136/100 36		206 223	507.5	225/187 38		B.M. 35 See F.B. S.R.-6 Pg.31 for Desc.
Total	+2604.193		575		-2440 041		576		
	-2440.041						575		
Diff in Elev	+164.152						1151 M		
B.M. 35									

FIGURE 59

LEVEL NOTES
FORKED-HAIR OPTICAL MICROMETER INSTRUMENT
See Section 6—4

16

P.C. & ℞ - M. Ratner
℞ - B. Beaudine

June 10, 1961
Weather - Cool & Windy
Inst. Zeiss 1234 - ℓ Lenker Topo

TOPO CONTROL LEVELS

STA	SET	READ	ELEV.	REF	DESCR.
STA	0.00		100.00	Assumed	Top of curb @ L&T Prod S.W.Cor Lot 16
TP	6.41	6.41	96.41		Top of rock
TP	1.82	1.82	91.82		Top of rock
TP	5.71	5.71	85.71	Pg.15	Top of 2"Ø.D.I.P.@ N.W.Cor Lot 16
TP	6.93	6.93	86.93	Pg.15	Top of 2"Ø.D.I.P @ N.E. Cor Lot 16
TP	0.36	0.36	90.36		Top of rock
TP	4.17	4.17	94.17	Pg.15	Top of 2"x2" stake Photo Setup "A"
TP	8.32	8.32	98.32	Pg.15	Top of curb @ L&T Prod S.E Cor Lot 16
TP		0.02	100.02	Assumed 100.00	Top of curb @ L&T.Prod SW Cor Lot 16

FIGURE 60

LEVEL NOTES
SINGLE WIRE, LENKER ROD
See Section 6-4

BEAMAN ARC LEVELS FOR TOPO CONTROL - OVER SADDLE BTWN BM's 6 & 8

June 5, 1958
Weather - Hot & still
Inst - K & E Transit with Beaman Arc

T.- B. Sibberts
φ.- B. Campbell
▯ - C. White #3666   28

Sta.	BS/FS	Dist.	Arc	Rod	Rod	Diff.	H.I.	Elev.	Adj. Elev.	Descriptions
	B.S.	150	38	⊕18.00	+8.15	+26.15	227.26	201.11	201.11	B.M. 6, Conc. Mon & Br. Cap (See pg. 22)
TP	F.S.	165	56	+9.90	-2.12	+7.78		235.04 / +07	235.11	Topo Control / T.B.M. #6-1, Set in 2 flush
	B.S.	134	40	⊕13.40	+12.36	+25.76	260.80			
TP	F.S.	177	55	+8.85	-1.71	+7.14		267.94 / +14	268.08	" #6-2 "
	B.S.	115	50	0	+10.31	+10.31	278.25			
	⓪ F.S.	130	66	+20.80	-2.03	+18.77				
TP	② F.S.	131	68	+23.58	-4.89	+18.69	⊗18.73	296.98 / +21	297.19	" #6-3 "
	B.S.	185	53	⊖5.53	+1.20	-4.35	292.63			
TP	F.S.	142	45	-7.10	-10.75	-17.85		274.78 / +28	275.06	" #6-4 "
	B.S.	171	53	⊖5.13	+1.50	-3.63	271.15			
TP	F.S.	138	41	-12.42	-9.82	-22.24		248.91 / +35	249.26	" #6-5 "
	B.S.	155	60	⊖15.50	+2.21	-13.29	235.62			
TP	F.S.	180	40	-18.00	-12.71	-30.71		204.91 / +42	205.33	" #6-6 "
	B.S.	147	61	⊖16.17	+1.71	-14.46	190.45			
	F.S.	138	42	-11.04	-8.79	-19.83		170.62 ≃ / +49	171.11	B.M. #8, Boat Spk. in P.P. 16 #4789E (See pg. 23)

171.11
−170.62

Error Of Closure = 0.49

$$\frac{0.49}{7 \; Turns} = 0.07 \; Per \; Turn$$

FIGURE 61

LEVEL NOTES
BEAMAN STADIA ARC
See Section 6—4

5

ALTIMETER OBSERVATIONS
WING PT. PHOTO CONTROL

Job # 4-1065
Flight line 6

Jan. 26, 1953
Observer - A. Smith
Cold & still

Station	Hour	Inst. Reading	Temp °F	Diff. of Readings	Mean Temp.	Temp. Adjustment % Incr.	Temp. Adjustment Total	Prem. Adj.	Lin Adj	Elevation Adj.	Elevation Actual
B.M.-SP7 (See Bk 7-21)	9:25	1305	38°		38.5°						1305.0
PP 6-A See Pht 6-2 Dscrp	9:32	1406	39°	+101	38.5°	-2.3	-2.3	1403.7	+0.4	1404	
PP6-B See Pht 6-2 Dscrp	9:45	1671	40°	+265	39.5°	-5.4	-7.7	1663.3	+1.2	1665	
PP6-C See Pht 6-2 Dscrp	9:56	1296	43°	-375	41.5°	+6.7	-1.0	1295.0	+1.9	1297	
PP6-D See Pht 6-4 Dscrp	10:12	1285	45°	-11	44°	+0.1	-0.9	1284.1	+2.9	1287	
PP6-E See Pht 6-4 Dscrp	10:20	1501	47°	+216	46°	-1.7	-2.6	1498.6	+3.3	1502	
PP6-F See Pht 6-6 Dscrp	10:43	1707	49°	+206	48°	-0.8	-3.4	1703.6	+4.8	1708	
PP6-G See Pht 6-6 Dscrp	11:08	1535	53°	-172	51°	+0.4	-3.0	1532.0	+6.3	1538	
B.M.-SP8 (See Bk 7-22)	11:25	1315	55°	-220	54°	+1.7	-1.3	1313.7	+7.9		1320.6

(Cont on pg 6)

FIGURE 62

LEVEL NOTES
ALTIMETER LEVELS
See Section 6-4

PROFILE – FROM S.W.COR. PARKING LOT TO EXISTING 24" DRAIN LINE

Sta.	+	H.I.	-	Elev's T.P.s & BM's Profile	Profile	Description
BM.#42			—	438.21		Highest point of R.R. Spur
	3.71	441.92				R.Pole #1234F Ref Page 27
TP	5.10	441.40	5.62	436.30		Metal turning pin
0+00			5.20		436.20	Spk & Tin @ Low point in pavement / S.W.Cor. Parking Lot
0+50			5.3		436.1	Ground
1+00			5.5		435.9	"
1+29			5.7		435.7	Top of ditch
1+33			7.3		434.1	" Bot " "
1+36			7.4		434.0	" " "
1+40			5.6		435.8	" Top " "
1+50			5.9		435.5	" " "
2+00			6.1		435.3	" " "
2+37			9.5		431.9	Top of 4" Cast Iron pipe line crossing
2+37			6.4		435.0	Ground at pipe line
2+50			6.6		434.8	Ground
TP 2+66.75	4.31	439.34	6.37	435.03	434.8	Top of 2x2 stake at angle point, ground
3+00			5.0		434.3	0.2' below top of stake / Ground
3+27.2			5.2		434.1	Top of curb
3+27.3			5.7		433.6	Gutter line
TP 3+45.2			5.31	434.03		Top of Spk & Tin set flush with pavement & R.R.Rd.

+13.12
17.30
E = 13.12

(Cont. on Page 37)

(See Pgs 30 to 35 for horizontal location and stake out of route)

Oct 30 1949
Cool & Sunny
π – A. White
φ – B. Jones
Chk ∅ – R. Smith

Wild Level #7265
+13.12
-17.30
- 4.18
438.21
434.03

FIGURE 63

PROFILE NOTES WITH PHILADELPHIA ROD

See Section 6-5

PROFILE - FROM S.W.COR. PARKING LOT TO EXISTING 24" DRAIN LINE

(See Pgs 30 to 35 for horizontal location and stake out of route)

Oct 20, 1949   Cool & Sunny
π - A.White
φ - B.Jones
Ch. & φφ - R.Smith
Wild Level #7265
Lenker Tape Rod #3

Sta.	Set	Read	Elev's T.P.'s & B.M.'s	Profile	Notes
B.M.#42	8.21		438.21		Highest point of R.R. Spk in P.Pole #1234 E. Ref. Page 27
T.P.	6.30	6.30	436.30	436.20	Metal turning pin
0+00				436.20	Spk & Tin @ low point in pavement S.W.Cor. Parking Lot
0+50				436.1	Ground
1+00				435.9	"
1+29				435.7	Top of ditch bank
1+33				434.1	Bot. " " "
1+36				434.0	" " " "
1+40				435.8	Top " " "
1+50				435.5	" " " "
2+00				435.3	" " " "
2+37				431.9	Top of 4" Cast Iron pipe line crossing
2+37				435.0	Ground at pipe line
2+50				434.8	Ground
T.P. 2+66.15	5.03	5.03	435.03	434.8	Top of 2x2 stake at angle point, ground 0.2' below top of stake
3+00				434.3	Ground
3+27.2				434.1	Top of curb
3+27.3				433.6	Gutter line
P 3+45.2		4.03	434.03		Top of Spk & Tin set flush with pavement & R&R Rd.

(Cont. on Page 37)

FIGURE 64

PROFILE NOTES WITH LENKER ROD
See Section 6-5

56

PROFILE – PIPE LINE – ALT. ROUTE "A"

Route previously staked & €'s stationed

Dec. 20, 1959
Cold & light snow
Inst - K&E with Beaman Arc #366

Ref. Job 4-613. Book 4 Pgs 30/45

P.C.#00 - D. White
Π - B. Jones
θ - C. Smith

STA.	B.S.	H.I.	Diff. El. +/-	Elev.	S.I.	Beaman Vert. Horz		Prod. Horz	Prod. Vert.	Rod	REMARKS
Π at 0+00 = 341+50	5.11	346.24		341.13							Top 1x2 Sta. 341+50 / T.B.M. (Book 4 Pg 23)
0+49			-3.1	343.1	49	50	100	49	0	-3.1	Ground
1+02			+1.6	347.8	102	52	99.8	102	+2.0	-0.4	Bag. rock outcrop
			+4.2	350.4	163	53	99.7	163	+4.9	-0.7	Pt. on "
			+7.5	353.7	214	54	99.6	213	+8.6	-1.1	End "
			+10.70	356.94	265	55	99.5	264	+13.25	-2.55	Top of slope
Π at 3+00 on line	5.23	362.17									
3+00				357.3	0	50	100	–	–	-4.9	Ground
3+51				358.1	51	50	100	51	–	-4.1	Ground
4+04				358.0	104	50	100	104	–	-4.2	Top of dry wash
4+13				351.5	113	50	100	113	–	-10.7	Bot "
4+26				351.7	126	50	100	126	–	-10.5	"
4+55				357.9	195	50	100	135	–	-4.3	Top "
4+82				353.3	182	50	100	182	–	-8.9	€ 8' dirt road
			-17.0	345.2	247	48	99.8	247	-4.9	-12.1	Ground
			-19.01	343.16	275	48	99.8	294	-5.90	-13.11	Top 1x2 stake flush / Lath mkd 5+94.50 €
Π at 6+00 on line	4.81	347.97									
6+55				341.8	55	50	100	55	–	-6.2	Ground
7+12				338.9	112	50	100	112	–	-9.1	"

(Cont. on pg 57)

FIGURE 65

PROFILE NOTES WITH BEAMAN STADIA ARC
See Section 6–5

CROSS SECTIONS   P₂ LINE

Oct. 10, 1960
Weather - Cool & clear
Inst. - Zeiss Opton #2345

T. & tℓ - D. Smith
Ch - B. Jones
ℓ - C. White

30

Sta.	+	H.I.	-	Elev.	Remarks
(Cont. on pg 31)					
3+50				6.8	
3+00				6.2	Radial →
2+85.15 B.C.				5.6	Radial →
2+50				4.8	
P#1	4.21	78.60	10.11	74.59	
2+00				8.1	
←					
1+50				7.5	
1+00				6.3	
0+50				5.7	
0+00				5.1	
B.M.#3	5.17	84.50		79.33	

Cross-section shots — reading / offset distance (All shots taken on natural ground except as noted):

Sta.	Left	ℓ	Right
3+50	8.7/75, 8.2/50, 7.7/25, 7.1/9	6.8	6.4/17, 6.0/25, 5.5/50, 4.9/75
3+00	8.1/75, 7.6/50, 7.1/25, 6.8/10	6.2	5.8/16, 5.1/25, 4.6/50, 4.1/75
2+85.15 B.C.	7.9/75, 7.4/50, 6.7/25, 6.1/10	5.6	5.1/14, 4.7/25, 4.1/50, 3.6/75
2+50	7.1/75, 6.3/50, 5.8/25, 5.1/10	4.8	4.4/16, 3.9/25, 3.6/50, 3.1/75
P#1	Metal turning pin		
2+00	9.8/75, 9.3/50, 8.8/25, 8.5/10	8.1	7.6/15, 7.2/25, 6.2/50, 6.1/75
←			6.5/50, 6.1/75
1+50	9.5/75, 8.8/50, 8.1/25, 7.8/16	7.5	7.1/16.3 E.Rd, 7.2/25, 7.1/25.7 E.Rd, 5.1/50, 4.7/75
1+00	8.5/75, 7.9/50, 7.3/25, 6.6/10	6.3	5.9/8, 5.5/25, 5.1/50, 4.7/75
0+50	7.9/75, 7.4/50, 6.9/25, 6.3/10	5.7	5.2/10, 4.8/25, 4.5/50, 4.0/75
0+00	7.1/75, 6.6/50, 6.2/25, 5.6/7	5.1	4.6/10, 4.3/25, 4.0/50, 3.7/75

See level Bk. 636 Pg. 59 for Descr.

FIGURE 66

ROUTE CROSS SECTIONS
See Section 6-6

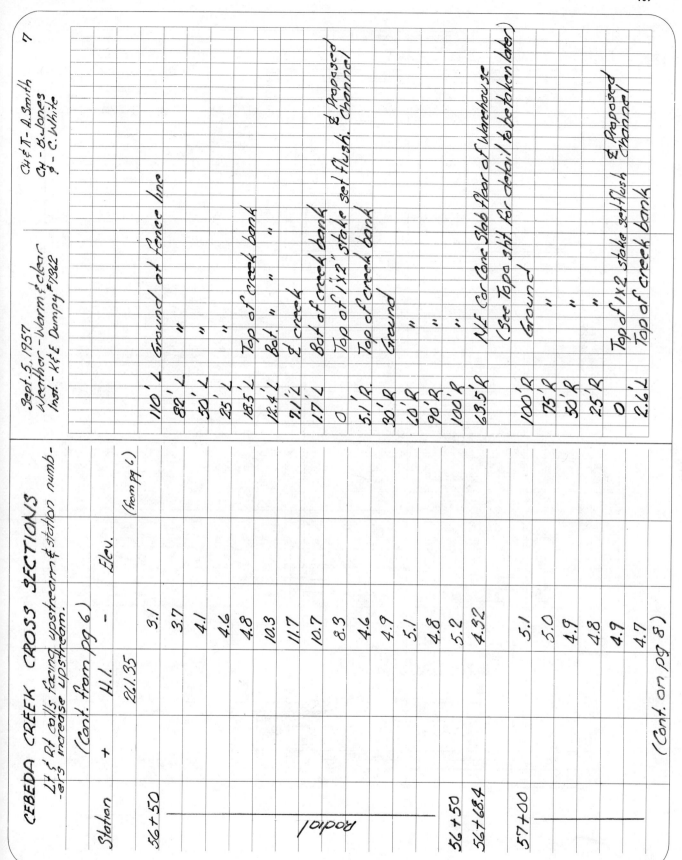

FIGURE 67

CROSS SECTIONS IN PROFILE FORM
See Section 6-6

110

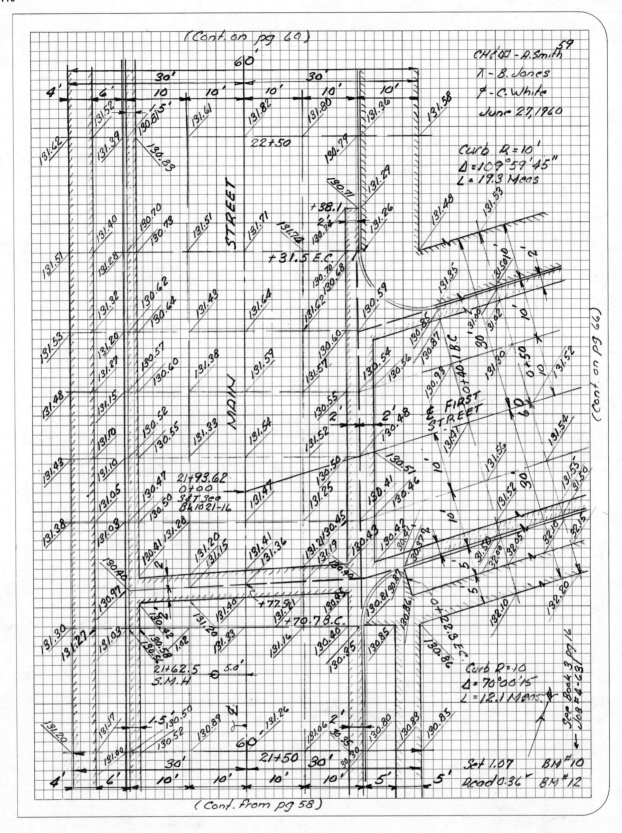

FIGURE 68

CROSS SECTIONS IN SKETCH FORM

See Section 6-6

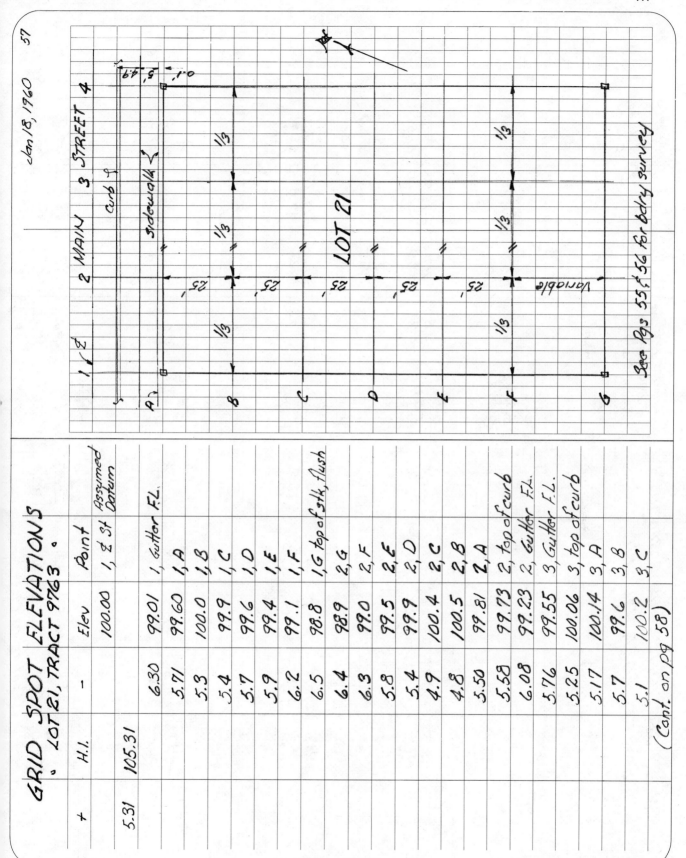

GRID SPOT ELEVATIONS
LOT 21, TRACT 9763

+	H.I.	−	Elev	Point
5.31	105.31		100.00	1, ₵ St Assumed Datum
		6.30	99.01	1, Gutter F.L.
		5.71	99.60	1,A
		5.3	100.0	1,B
		5.4	99.9	1,C
		5.7	99.6	1,D
		5.9	99.4	1,E
		6.2	99.1	1,F
		6.5	98.8	1,G top of stk, flush
		6.4	98.9	2,G
		6.3	99.0	2,F
		5.8	99.5	2,E
		5.4	99.9	2,D
		4.9	100.4	2,C
		4.8	100.5	2,B
		5.50	99.81	2,A
		5.58	99.73	2, top of curb
		6.08	99.23	2, Gutter F.L.
		5.76	99.55	3, Gutter F.L.
		5.25	100.06	3, top of curb
		5.17	100.14	3,A
		5.7	99.6	3,B
		5.1	100.2	3,C

(Cont. on pg 58)

Jan. 18, 1960   *57*

LOT 21

See Pgs 55 & 56 for bdry survey

FIGURE 69

CROSS SECTIONS DESCRIPTIVE GRID METHOD

See Section 6-6

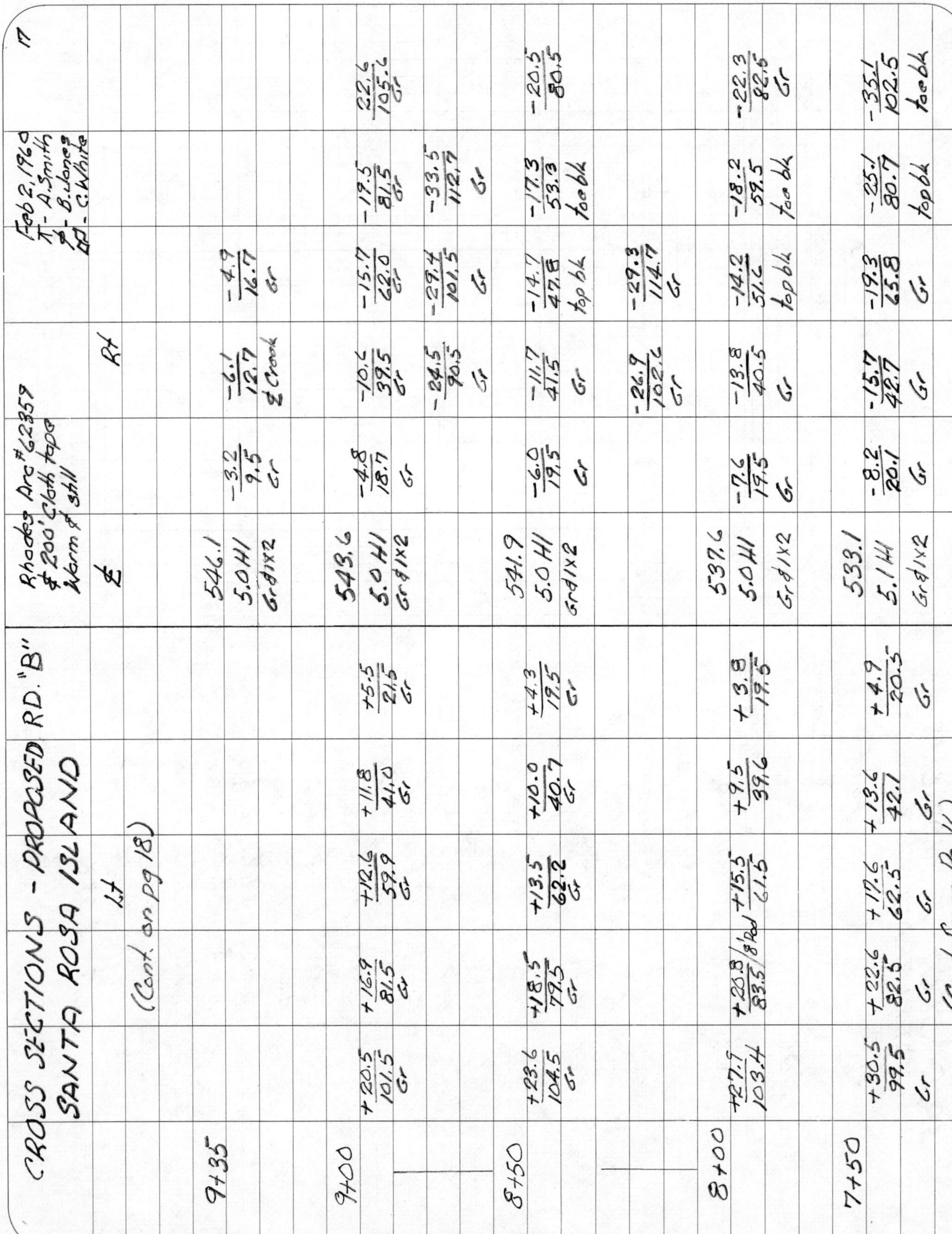

17

CROSS SECTIONS - PROPOSED RD. "B"
SANTA ROSA ISLAND

Feb 2, 1960
π - A. Smith
φ - B. Jones
∅ - C. White

Rhodes Arc #62359
℄ 200' Cloth tape
Warm & still

Lt				℄		Rt		

(Cont. on pg 18)

9+35

9+00
+20.5 / 101.5 / Gr +16.9 / 81.5 / Gr +12.6 / 59.9 / Gr +11.8 / 44.0 / Gr +5.5 / 21.5 / Gr | 546.1 / 5.0 HI / Gr 91x2 | -3.2 / 9.5 / Gr -6.1 / 12.7 / ℄ Creek -4.9 / 16.7 / Gr 22.6 / 105.6 / Gr

8+50
+23.6 / 104.5 / Gr +18.5 / 79.5 / Gr +13.5 / 62.2 / Gr +10.0 / 40.7 / Gr +4.3 / 19.5 / Gr | 543.6 / 5.0 HI / Gr 91x2 | -4.8 / 18.7 / Gr -10.6 / 39.5 / Gr -15.7 / 62.0 / Gr -19.5 / 81.5 / Gr

541.9 / 5.0 HI / Gr 91x2 -6.0 / 19.5 / Gr -24.5 / 90.5 / Gr -29.4 / 101.5 / Gr -33.5 / 112.7 / Gr

8+00
+27.5 / 103.4 / Gr +20.8 / 83.5 / (8 bd) / Gr +15.5 / 61.5 / Gr +9.5 / 39.6 / Gr +3.8 / 17.5 / Gr | 537.6 / 5.0 HI / Gr 91x2 -7.6 / 19.5 / Gr -13.8 / 40.5 / Gr -14.7 / 47.8 / top bk -17.3 / 53.3 / too bk -20.5 / 80.5 / Gr

-26.9 / 102.6 / Gr -29.3 / 114.7 / Gr

7+50
+30.5 / 99.5 / Gr +22.6 / 82.5 / Gr +17.6 / 62.5 / Gr +13.6 / 42.1 / Gr +4.9 / 20.5 / Gr | 533.1 / 5.1 HI / Gr 91x2 -8.2 / 20.1 / Gr -15.7 / 42.7 / Gr -19.3 / 25.8 / Gr -25.1 / 80.7 / top bk -35.1 / 102.5 / too bk

(Cont from pg 16)

FIGURE 70

CROSS SECTIONS USING RHODES ARC
See Section 6-6

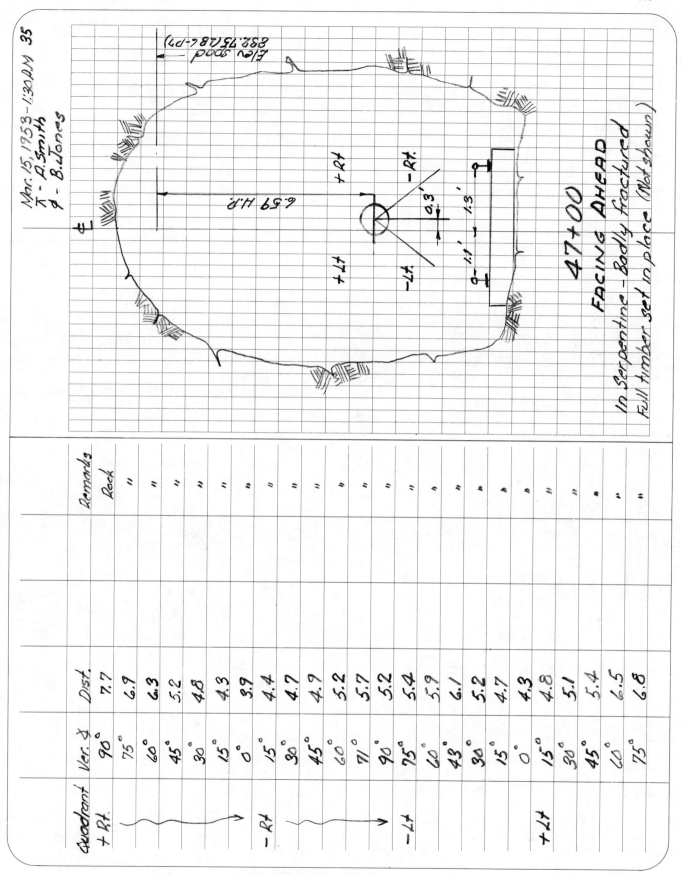

FIGURE 71

CROSS SECTIONS
TUNNEL SURVEY
See Section 6-6

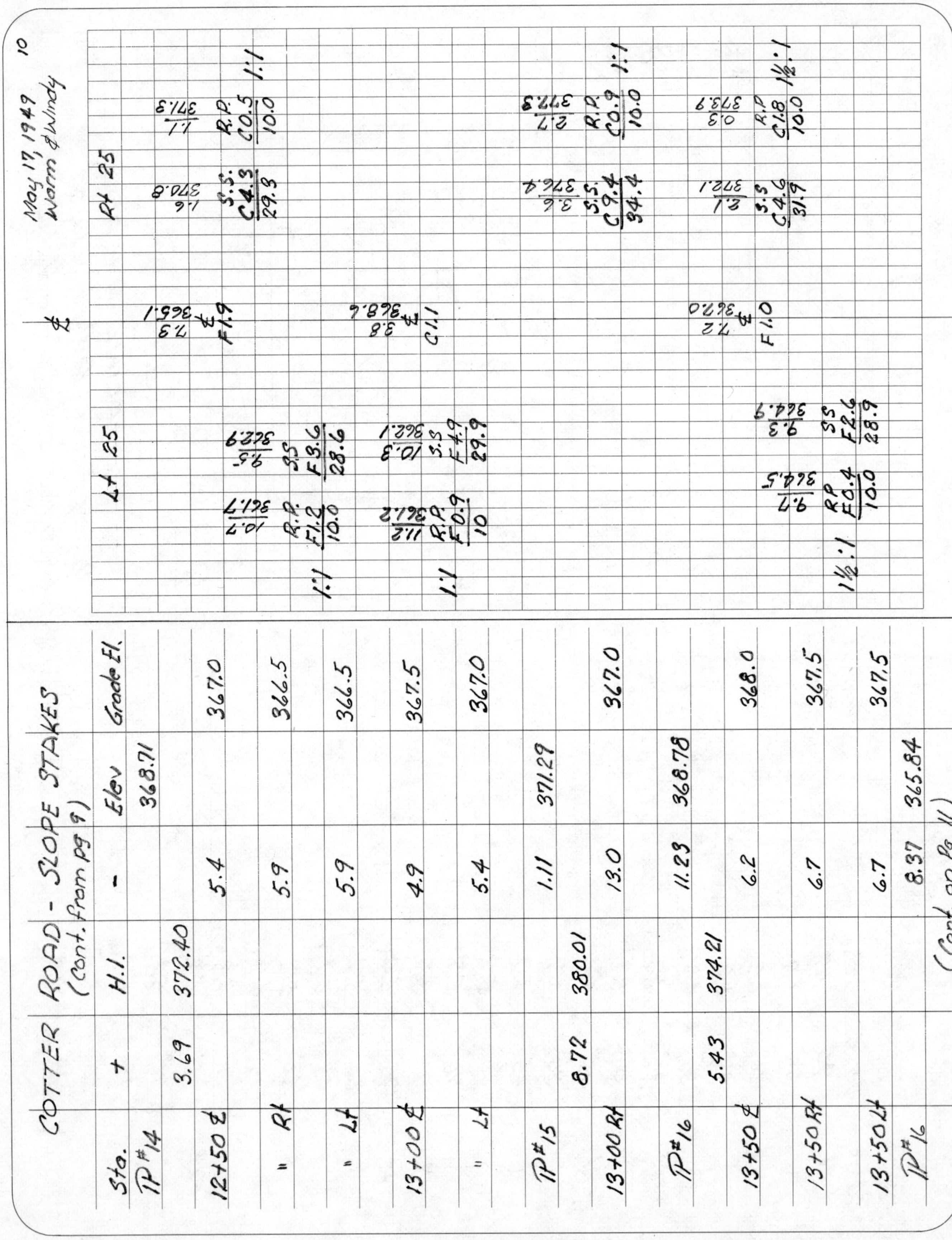

May 17, 1949 Warm & Windy
10
Rt 25
℄
Lt 25
1:1
1½:1

COTTER ROAD - SLOPE STAKES (Cont. from pg 9)

Sta.	+	H.I.	-	Elev	Grade El.
TP#14	3.69	372.40		368.71	
12+50 ℄			5.4		367.0
" Rt			5.9		366.5
" Lt			5.9		366.5
13+00 ℄			4.9		367.5
" Lt			5.4		367.0
TP#15	8.72	380.01	1.11	371.29	
13+00 Rt			13.0		367.0
TP#16	5.43	374.21	11.23	368.78	
13+50 ℄			6.2		368.0
13+50 Rt			6.7		367.5
13+50 Lt			6.7		367.5
TP#16			8.37	365.84	

(Cont. on Pg 11)

FIGURE 72

SLOPE STAKING NOTES
DESCRIPTIVE FORM
See Section 6–7

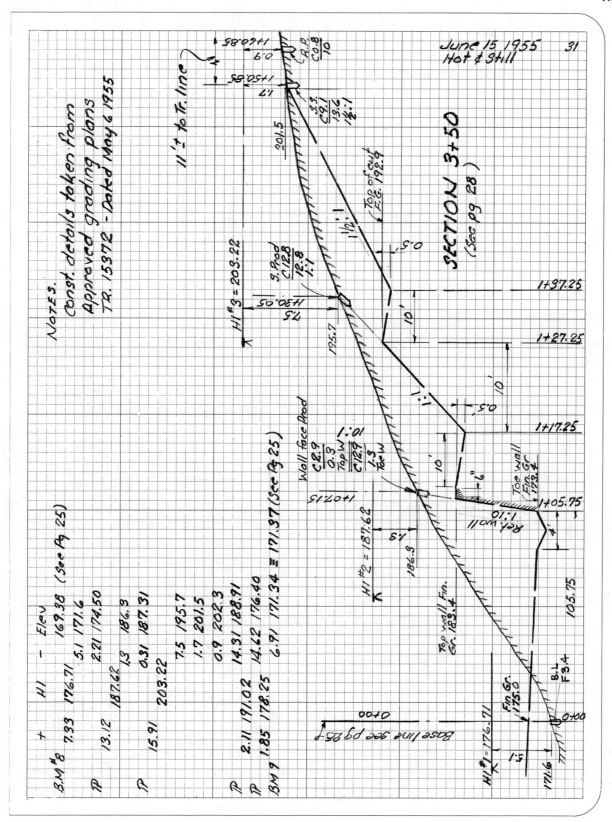

FIGURE 73

SLOPE STAKING NOTES
GRAPHICAL FORM
See Section 6-7

116

SLOPE STAKE

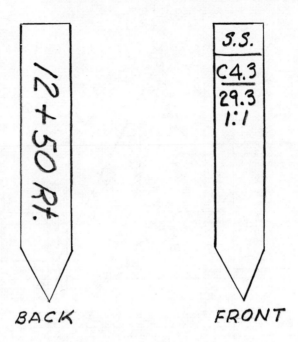

BACK   FRONT

REFERENCE STAKE FOR SLOPE STAKE

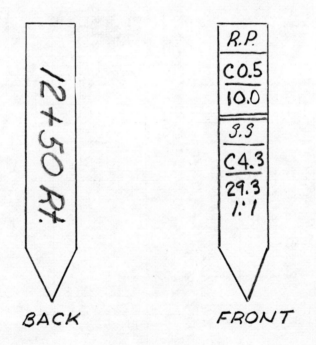

BACK   FRONT

FIGURE 74

SLOPE STAKING
MARKING GUARD STAKES
See Section 6-7

36

June 22, 1959
T - A. Smith
ϕ - B. Brown

Warm
Inst. - K&E Transit #5762

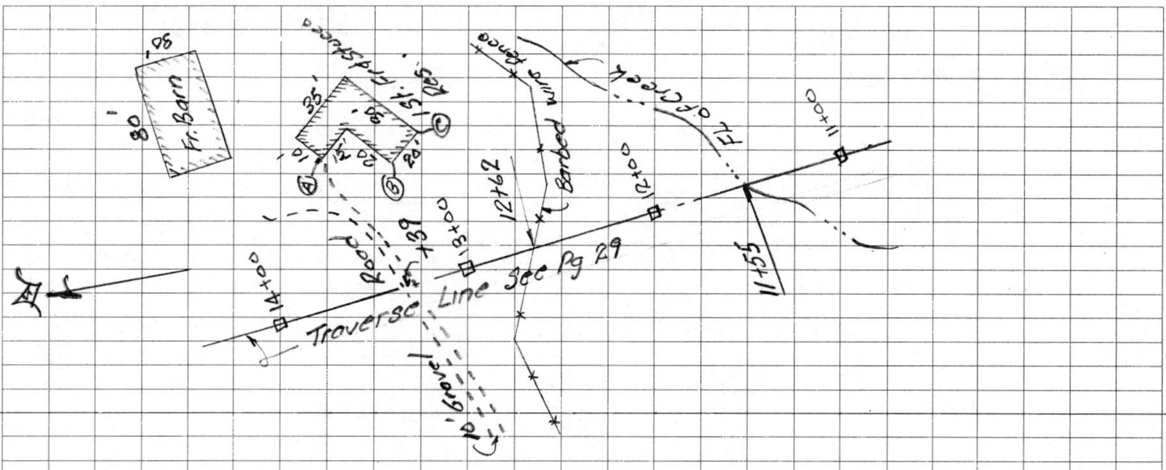

TRANSIT STADIA TOPO.
STATIONS 11+00 to 14+00

S.I./H.Dist	Hor ∠	Vert ∠	Rod	Ver Diff.	Elev.	Descrip.
T @ 13+00	0°00'	—	+5.1		320.61	14+00 (from Bk 6 Pg 21)
1.52/152	38°16	—	-6.1			NW Cor Barn
1.20/120	39°46	—	-6.0			SW " "
.67/67	42°06'	—	-5.0			Cor House Pt Ⓐ
.48/48	82°31'	—	-4.9			" " Pt Ⓑ
50/50	95°48	—	-4.9			" " Pt Ⓒ
108/108	163°16	—	-5.9			Pt. on fence line
89/89	116°07	—	-6.8			⑤ Pt of "
60/60	171°03'	—	-7.7			" "
58/58	266°59'	—	-9.3			" "
105/105	267°33	—	-11.8			Pt on "
110	130°11	-0°16	-10.7			Creek bank
118	133°08	-0°26	-11.5			FL creek
123	137°46	-0°17'	-10.7			Creek bank
117	158°12	-0°22	-12.8			" "
123	161°17	-0°35'	-13.1			FL creek
133	162°45	-0°22'	-12.1			Creek bank
165	197°30	-1°35'	-10.7			" "
171	199°59'	-1°47'	-12.8			FL creek
178	202°31	-1°40'	-11.7			Creek bank
230	220°15'	-1°15'	-10.8			Ground
278	281°07'	-1°18'	-11.7			" "
217	253°17'	-2°10'	-12.8			" "

Angles clockwise - H.I. = 325.7

(Cont. on pg 37)

FIGURE 75
TOPOGRAPHY
TRANSIT STADIA NOTES
See Section 6—8

118

July 15, 1959
Hot & light wind

π — A. Smith   45
⊠ — B. Jones
9 — C.White & D.White

PLANE TABLE SHEET #9

DIST.	ARC	PROD	ROD	DIFF.EL	HI	ELEV.	NOTES
π @ Δ #15		+6.36			365.59	359.23	T.B.M. #12
315	56	+18.9	3.4	+15.5		381.1	Ground
370	56	+22.2	6.5	+15.7		381.3	"
283	50	−	8.1			357.5	"
310	50	−	6.2			359.4	"
410	48	−8.2	10.0	−18.2		347.4	"
370	50	−	9.1			356.5	"
310	52	+6.2	6.2	0		365.6	"
290	50	−	7.1			358.5	E. 20' Rd
304	50	−	6.9			358.7	₵ 20' "
318	50	−	7.0			358.6	W 20' "
π @ T.P. #7		+3.6			344.31	340.71	T.P. #7
75	45	−3.7	10.2	−13.9		330.4	Top bank
86	40	−8.6	11.2	−19.8		324.5 ~~370.5~~	Toe bank
230	45	−11.5	9.5	−21.0		323.3	Top bluff
245	42	−19.6	12.5	−32.1		312.2	Toe bluff
323	45	−16.2	10.2	−26.4		317.9	Ground
375	50	−	2.8			341.5	"
304	50	−	4.3			340.0	"
236	50	−	7.6			336.7	"
436	50	−	9.22			335.09	T.P.

(cont. on pg 46)

FIGURE 76

TOPOGRAPHY
PLANE TABLE NOTES
See Section 6−8

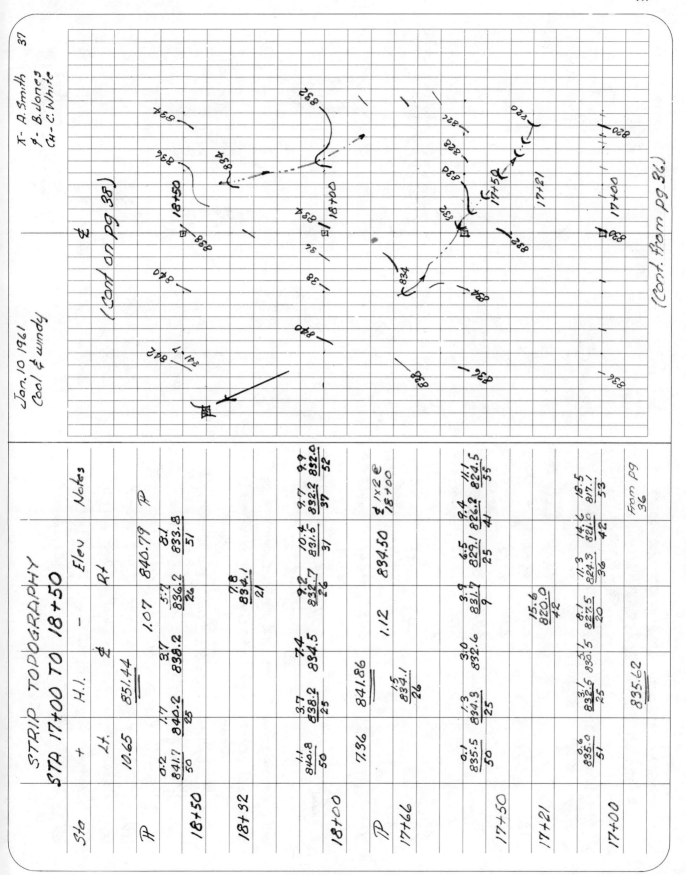

FIGURE 77

TOPOGRAPHY NOTES
CROSS SECTION AND SKETCH FORM
See Section 6-8

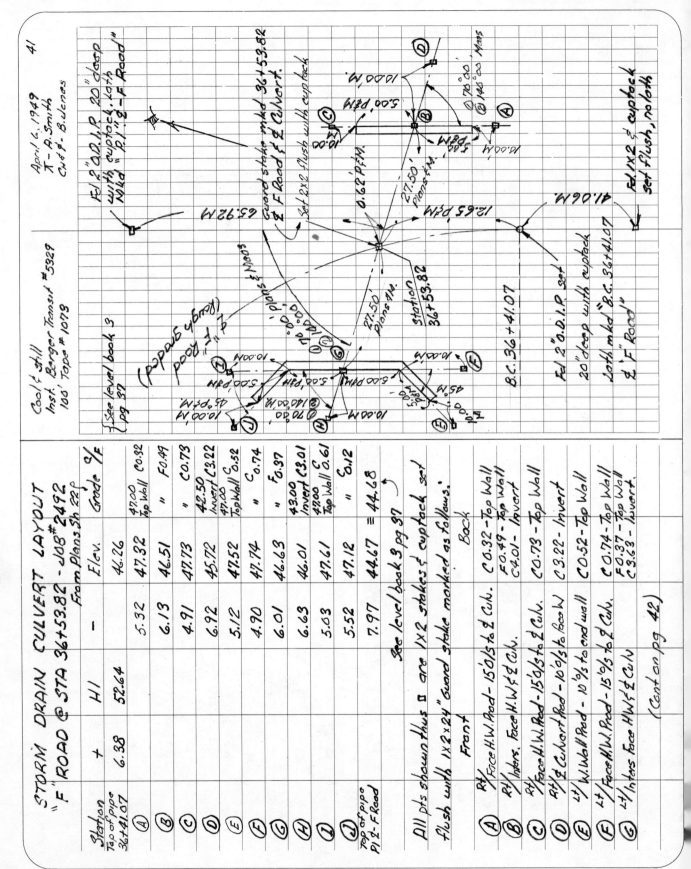

FIGURE 78

CONSTRUCTION NOTES
CULVERT LAYOUT
See Section 6-9

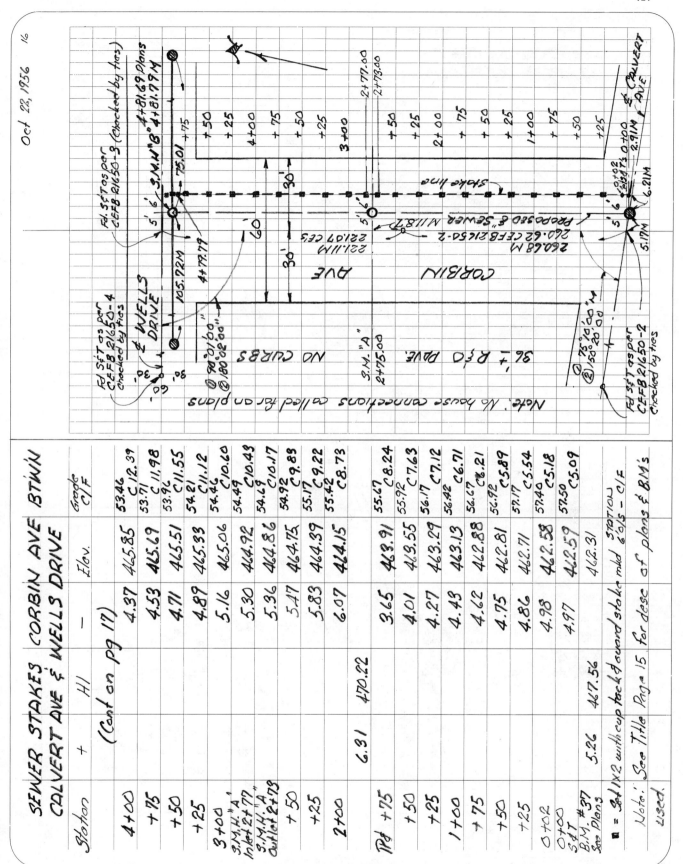

FIGURE 79

CONSTRUCTION NOTES
SEWER STAKES
See Section 6-9

122

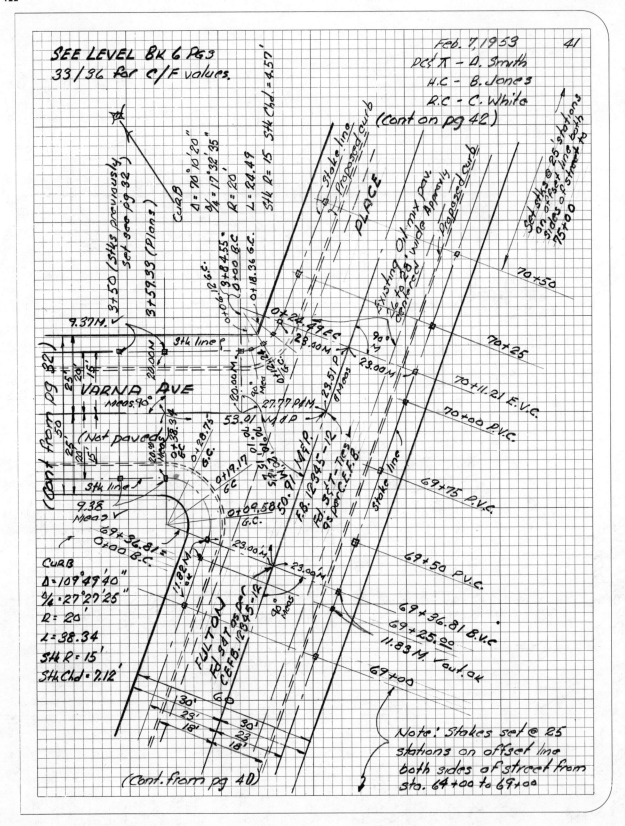

FIGURE 80

CONSTRUCTION NOTES
CURB STAKES
See Section 6-9

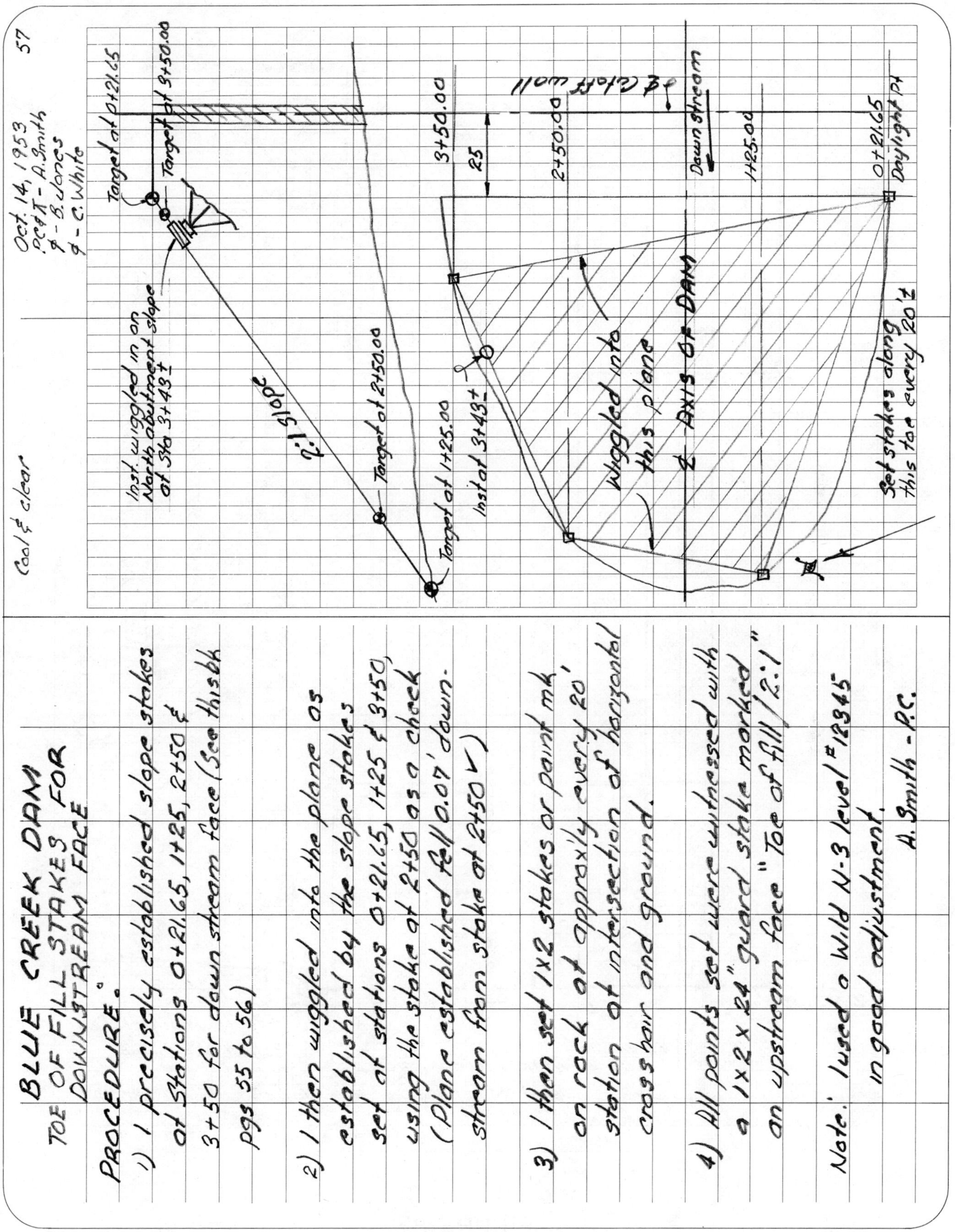
BLUE CREEK DAM
TOE OF FILL STAKES FOR DOWNSTREAM FACE
PROCEDURE:
1) I precisely established slope stakes at Stations 0+21.65, 1+25, 2+50 & 3+50 for down stream face (See this bk pgs 55 to 56)
2) I then wiggled into the plane as established by the slope stakes set at stations 0+21.65, 1+25 & 3+50, using the stake at 2+50 as a check (Plane established fell 0.07' down-stream from stake at 2+50 ✓)
3) I then set 1×2 stakes or paint mk on rock at approx'ly every 20' station at intersection of horizontal crosshair and ground.
4) All points set were witnessed with a 1×2×24" guard stake marked on upstream face "Toe of fill / 2:1"
Note: I used a Wild N-3 level #12345 in good adjustment.
A. Smith - P.C.
Cool & clear
Oct. 14, 1953
PC & ⊼ - A. Smith
φ - B. Jones
φ - C. White
57
Target at 0+21.65
Inst. wiggled in on North abutment slope at Sta 3+43±
Target at 3+50.00
2:1 slope
Target at 2+50.00
Target at 1+25.00
3+50.00
Inst at 3+43±
25
2+50.00
Wiggled into this plane
℄ AXIS OF DAM
℄ cutoff wall
Down stream
1+25.00
0+21.65
Daylight Pt
Set stakes along this toe every 20'±

FIGURE 81

CONSTRUCTION NOTES
SLOPE STAKES FOR A DAM
See Section 6–9

124

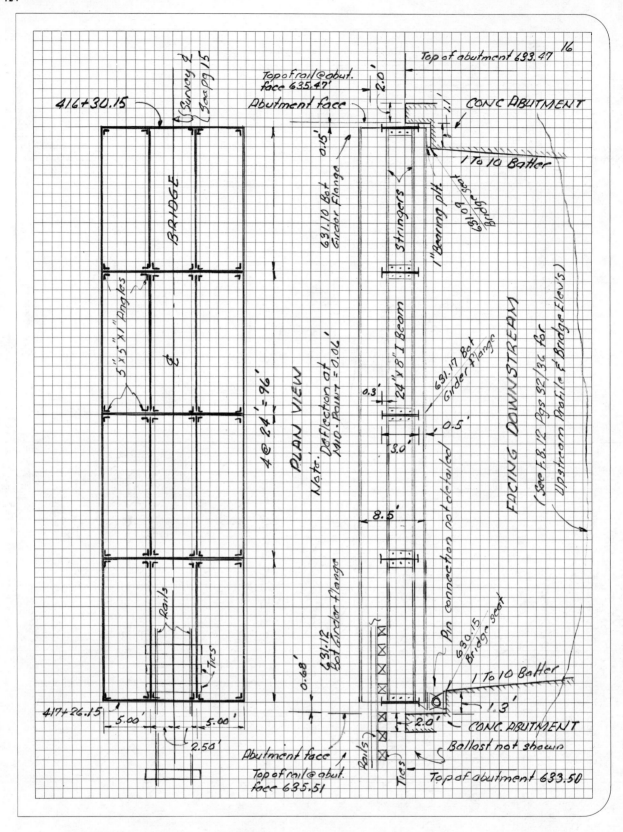

FIGURE 82

AS–BUILT NOTES
BRIDGE DETAIL
See Section 6–10

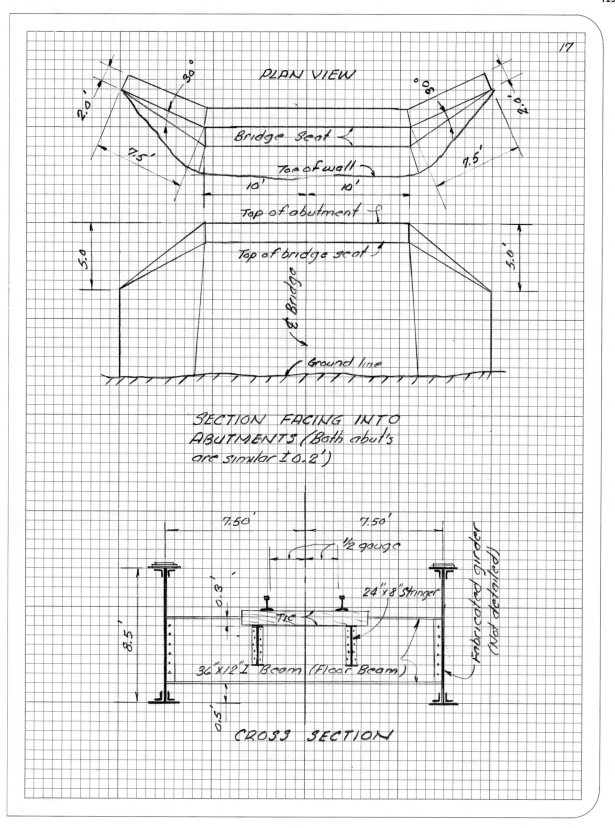

FIGURE 83

AS-BUILT NOTES
BRIDGE DETAIL
See Section 6-10

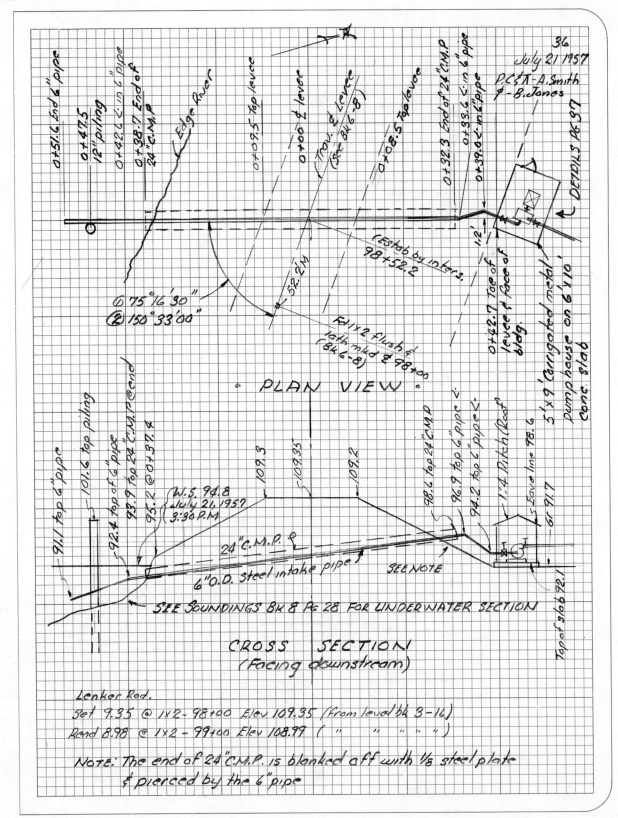

FIGURE 84

AS—BUILT NOTES
STRUCTURE SECTION
See Section 6—10

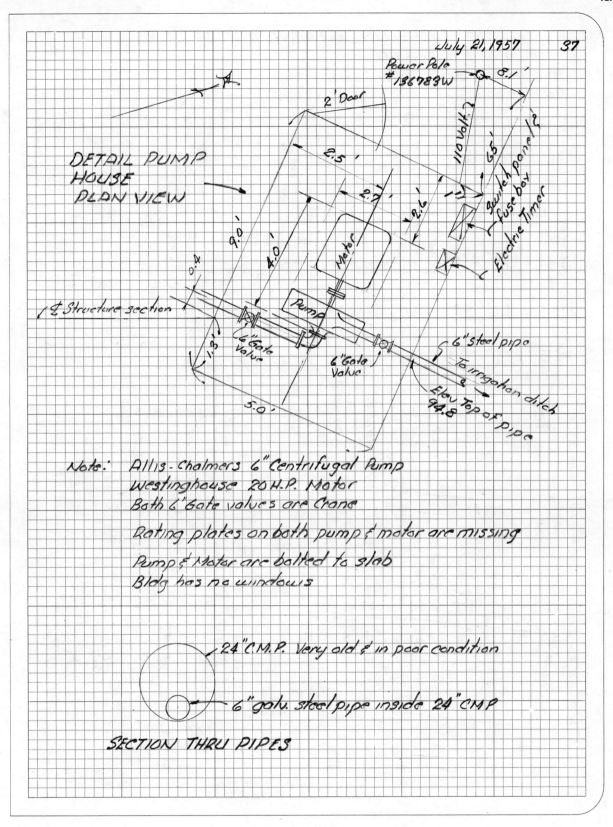

Power Pole
#136783W

110 Volt.?

Switch panel 1/3
fuse box
Electric Timer

DETAIL PUMP
HOUSE
PLAN VIEW

2' Door

2.5'

2.7'

9.0'

4.0'

0.4

℄ Structure section

1.3'

6" Gate
Valve

Pump

Motor

2.6'

6.5

8.1'

6" Gate
Valve

5.0'

6" steel pipe

To irrigation ditch

Elev Top of Pipe
94.8

Note: Allis-Chalmers 6" Centrifugal Pump
        Westinghouse 20 H.P. Motor
        Both 6" Gate valves are Crane

        Rating plates on both pump & motor are missing

        Pump & Motor are bolted to slab
        Bldg has no windows

24" C.M.P. Very old & in poor condition

6" galv. steel pipe inside 24" CMP

SECTION THRU PIPES

FIGURE 85

AS-BUILT NOTES
STRUCTURE SECTION
See Section 6-10

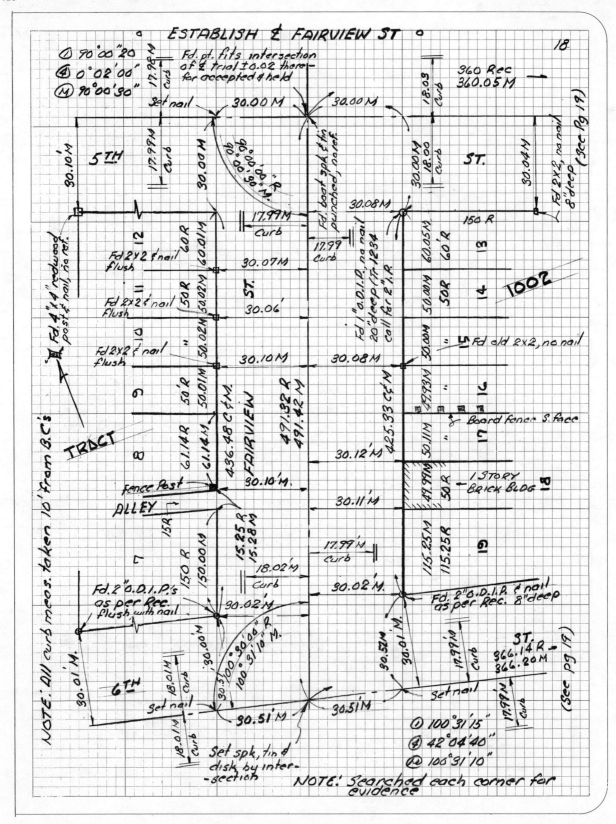

FIGURE 86

BOUNDARY SURVEY NOTES
ESTABLISHMENT OF CENTER LINE
See Section 6—11

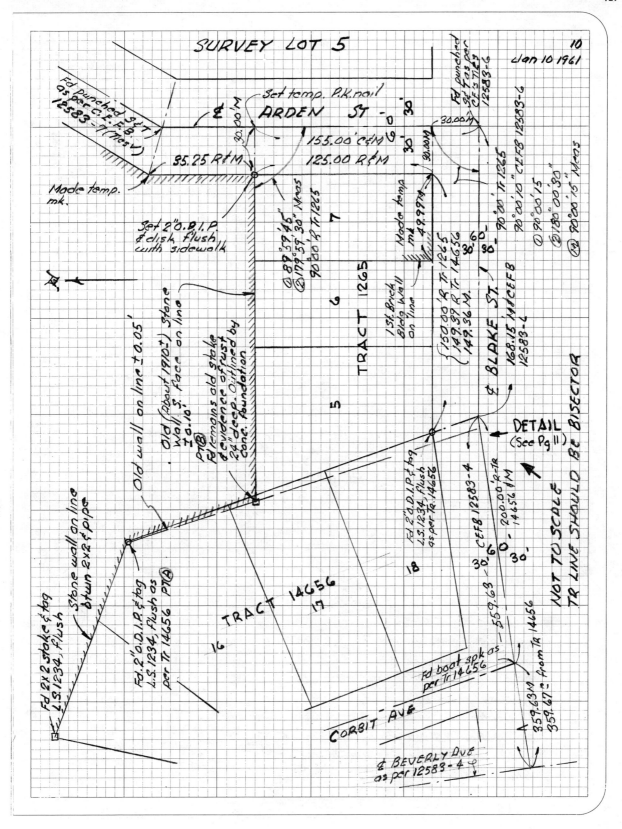

FIGURE 87

BOUNDARY SURVEY NOTES
LOT SURVEY
See Section 6-11

130

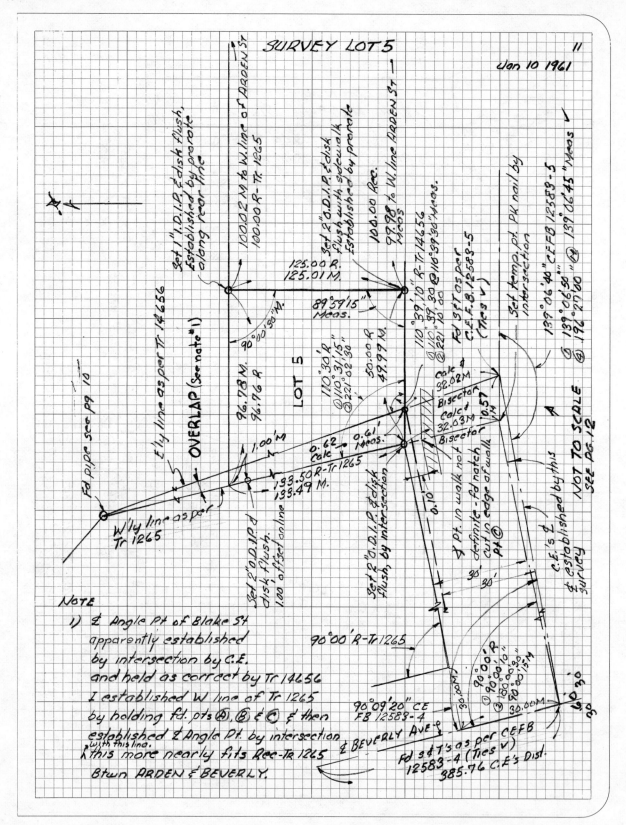

FIGURE 88

BOUNDARY SURVEY NOTES
LOT SURVEY
See Section 6-11

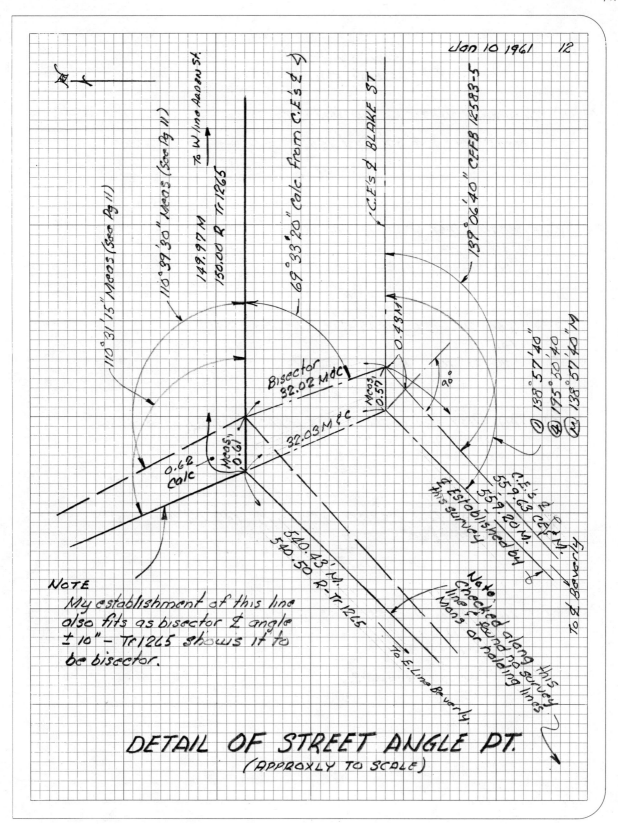

NOTE

My establishment of this line
also fits as bisector ∠ angle
± 10" - Tr 1265 shows it to
be bisector.

DETAIL OF STREET ANGLE PT.
(APPROXLY TO SCALE)

FIGURE 89

BOUNDARY SURVEY NOTES
LOT SURVEY
See Section 6-11

SURVEY LOT 5
FIELD CALC'S

Jan 10, 1961

① SLOPE REDUCTION ORDEN ST.

Meas Vert ∡ 4°30' Dist 155.48   Ex. sec = .00309

$$\begin{array}{r} 155.48 \\ -.48 \\ \hline 155.00 \end{array}$$

.309'	– 100
.154	– 50
.015	– 5
.001	– 0.5
.479	155.5

② BISECTOR DIST C.E's ∡ 139°06'40"

$$2)\ 40°53'20"$$
$$20°26'40" \quad Ex\ Sec = .0672$$
$$\times 30$$
$$\overline{2.016}$$

= 32.02

③ BISECTOR DIST. MEAS ∡ @ P.I. 138°57'40"

$$2)\ 41°02'20"$$
$$20°31'10" \quad Ex\ sec = .0677$$
$$\times 30$$
$$\overline{2.0310}$$

= 32.03

④ TAN. DIST FOR BISECTOR OF MEAS ∡ AT P.I. 138°57'40"

$$TAN.\ 20°31'10" = .3742$$
$$\times 30$$
$$\overline{11.2260}$$

⑤ TAN. DIST FOR BISECTOR C.E.'s ∡ 139°06'40"

$$TAN\ 20°26'40" = .3727$$
$$\times 30$$
$$\overline{11.1810}$$

⑥ DIST BTWN BISECTORS @ P.L.

$$\begin{array}{r} 11.23 \\ -11.18 \\ \hline 0.05 \\ +0.57\ Meas\ @\ ℄ \\ \hline 0.62' \end{array}$$

FIGURE 90

BOUNDARY SURVEY NOTES
FIELD CALCULATIONS
See Section 6–11

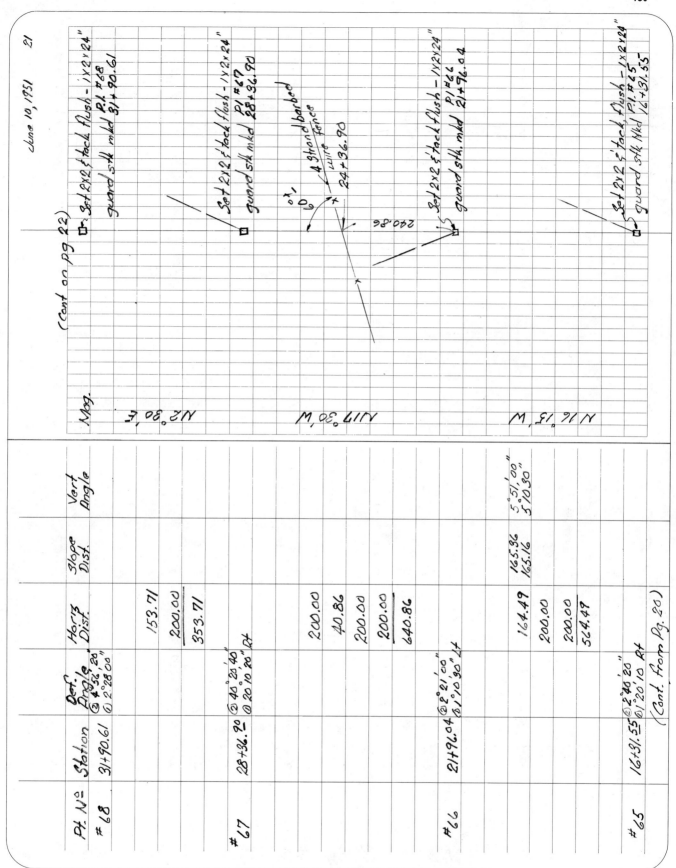

FIGURE 91

TRAVERSE NOTES
TRANSIT—TAPE
See Section 6—11

RECONNAISSANCE TRAVERSE
PROPOSED PIPE LINE

Dec. 10, 1945 — ℭ of tunnel clockwise — 55

Pt. occupied	S.I.	Horiz. Angle	Vert. Angle	Rod	Elev	Pt. sighted	
∠ #11	Rod @ T = 4.95		HI =				(Cont. from Pg 54)
	306	0°00'00"	+6°10'30"	4.95		∠ #10	
	485	①165°10'20"	-5°06'00"	4.95		∠ #12	Set 2x2⅝' tack flush, 1x2x24" guard stk mkd ∠ #12
		②336°20'30"					
Mag N25°15'W							
∠ #12	Rod @ T = 5.20		HI =				
	437	0°00'00"	+5°06'30"	5.20		∠ #11	
	316	①193°44.30'	-2°10'15"	5.20		∠ #13	Set 2x2⅝ tack flush, 1x2x24" guard stk mkd ∠ #13
		②27°33'00"					
Mag N11°30'W							
∠ #13	Rod @ T = 5.15		HI =				
	315	0°00'00"	+2°10'15"	5.15		∠ #12	
	473	①176°03'00"	-1°30'30"	5.15		∠ #14	Set 2x2⅝ tack flush, 1x2x24" guard stk mkd ∠ #14
		②352°06'00"					
Mag N15°30'W							
∠ #14	Rod @ T = 5.05		H.I. =				
	471	0°00'00"	+1°30'20"	5.05		∠ #13	
	287	①139°56'30"	-4°16'30"	5.05		∠ #15	Set 2x2⅝ tack flush, 1x2x24" guard stk mkd ∠ #15
		②279°53'00"					
Mag N55°30'W							
∠ #15	Rod @ T = 4.90		HI =				
	288	0°00'00"	+4°16'30"	4.90		∠ #14	
	306	①178°06'10"	-3°08'10"	12.85		∠ #16	Set 2x2⅝ tack flush, 1x2x24" guard stk mkd ∠ #16
		②356°12'30"					(Cont on Pg 56)
Mag N53°30'W							

FIGURE 92

TRAVERSE NOTES
STADIA
See Section 6-12

35

Jan 27, 1955
P.C. 6°CLO - A.Smith
π - B.Jones

Cool & Clear
Inst. Wild T-2 #36584
Subtense Bar #1365

INST. @ TRAV. STA. #65

Position	Pt. Obs.	Observation	Mean D&R	Direction	Angle Across Bar	Dist from tables
#1	Sta #64	0°00'16" 180°00'22"	0°00'19"	0°00'00"		
	Bar	181°29'19" 1°29'17"	181°29'18	181°28'59"		
	Sta #66	182°09'30" 2°09'32"	182°09'31	182°09'12"		
	Bar	182°49'41" 2°49'43"	182°49'42	182°49'23"	1°20'24"	280.56
#2	Sta #64	31°20'32" 211°20'30	31°20'31	0°00'00"		
	Bar	212°49'18 32°49'20	212°49'19	181°28'48"		
	Sta #66	213°29'47" 33°29'45	213°29'46	182°09'15"		
	Bar	214°09'48 34°09'46	214°09'47	182°49'16"	1°20'28"	280.32

STA #65 - Fd T Bar & disk, flush, stamped
"STA #65 - Job #4-631"

(Cont. Pg 36)

FIGURE 93

TRAVERSE NOTES
SUBTENSE BAR
See Section 6-12

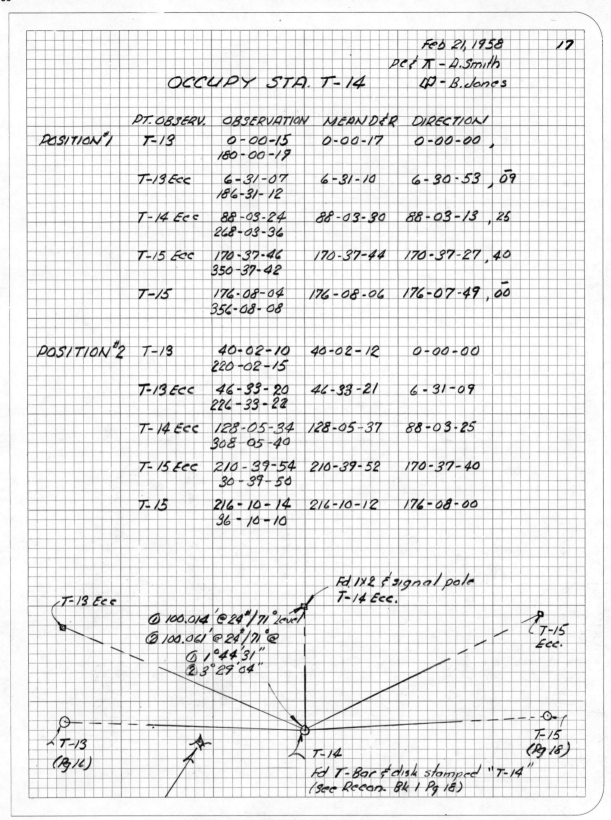

136

Feb 21, 1958   17

pct П – D. Smith
ΙΦ – B. Jones

OCCUPY STA. T-14

	PT. OBSERV.	OBSERVATION	MEANDER	DIRECTION
POSITION #1	T-13	0-00-15 180-00-19	0-00-17	0-00-00 ,
	T-13 Ecc	6-31-07 186-31-12	6-31-10	6-30-53 , 09̄
	T-14 Ecc	88-03-24 268-03-36	88-03-30	88-03-13 , 25
	T-15 Ecc	170-37-46 350-37-42	170-37-44	170-37-27 , 40
	T-15	176-08-04 356-08-08	176-08-06	176-07-49 , 00̄
POSITION #2	T-13	40-02-10 220-02-15	40-02-12	0-00-00
	T-13 Ecc	46-33-20 226-33-22	46-33-21	6-31-09
	T-14 Ecc	128-05-34 308-05-40	128-05-37	88-03-25
	T-15 Ecc	210-39-54 30-39-50	210-39-52	170-37-40
	T-15	216-10-14 36-10-10	216-10-12	176-08-00

T-13 Ecc

Fd 1×2 & signal pole
T-14 Ecc.

① 100.014' @ 24 #/71° level
② 100.061' @ 24 #/71° @
① 1° 44' 31"
② 3° 29' 04"

T-15
Ecc.

T-13
(Pg 16)

T-14

T-15
(Pg 18)

Fd T-Bar & disk stamped "T-14"
(see Recon. Bk I Pg 18)

FIGURE 94

TRAVERSE NOTES
SUBTENSE BASE
See Section 6-12

12

Sept. 3, 1951
P.C.§ ☼ - D.Smith
Π - B.Jones

Weather - Warm & Clear
Inst. - Berger #12345

Sept. 3, 1951
Watch checked 6:30AM P.S.T. @ W.U. 1m 21s slow
   "   "   5:10PM P.S.T.   "   "   1m 46s "
Latitude of 43+06.56 scaled from U.S.G.S.
Quad. Sht. φ = 34 14'12"

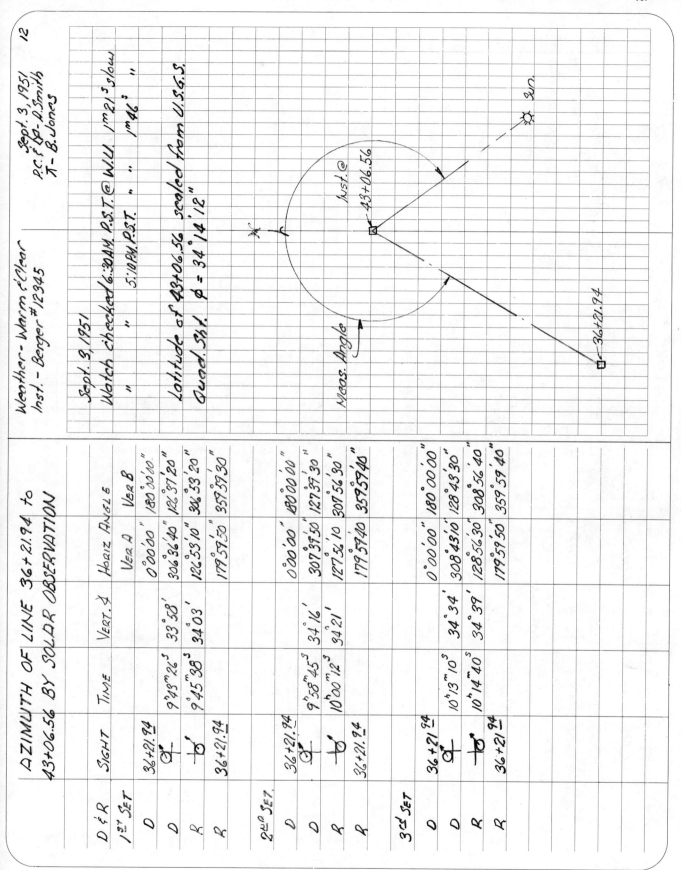

AZIMUTH OF LINE 36+21.94 to 43+06.56 BY SOLAR OBSERVATION

D & R	SIGHT	TIME	VERT. ∢	HORIZ ANGLE	
				VER A	VER B
1ST SET					
D	36+21.94			0°00'00"	180°00'00"
D	☼	9h 43m 26s	33°58'	306°36'40"	126°37'20"
R	☼	9h 45m 38s	34°03'	126°53'10"	306°53'20"
R	36+21.94			179°59'50"	359°59'30"
2ND SET					
D	36+21.94			0°00'00"	180°00'00"
D	☼	9h 58m 45s	34°16'	307°37'50"	127°37'30"
R	☼	10h 00m 12s	34°21'	127°56'10"	307°56'30"
R	36+21.94			179°59'40"	359°57'40"
3RD SET					
D	36+21.94			0°00'00"	180°00'00"
D	☼	10h 13m 10s	34°34'	308°43'10"	128°43'30"
R	☼	10h 14m 40s	34°39'	128°56'30"	308°56'40"
R	36+21.94			179°59'50"	359°59'40"

FIGURE 95

ASTRONOMIC NOTES
SOLAR OBSERVATION
See Section 6-12

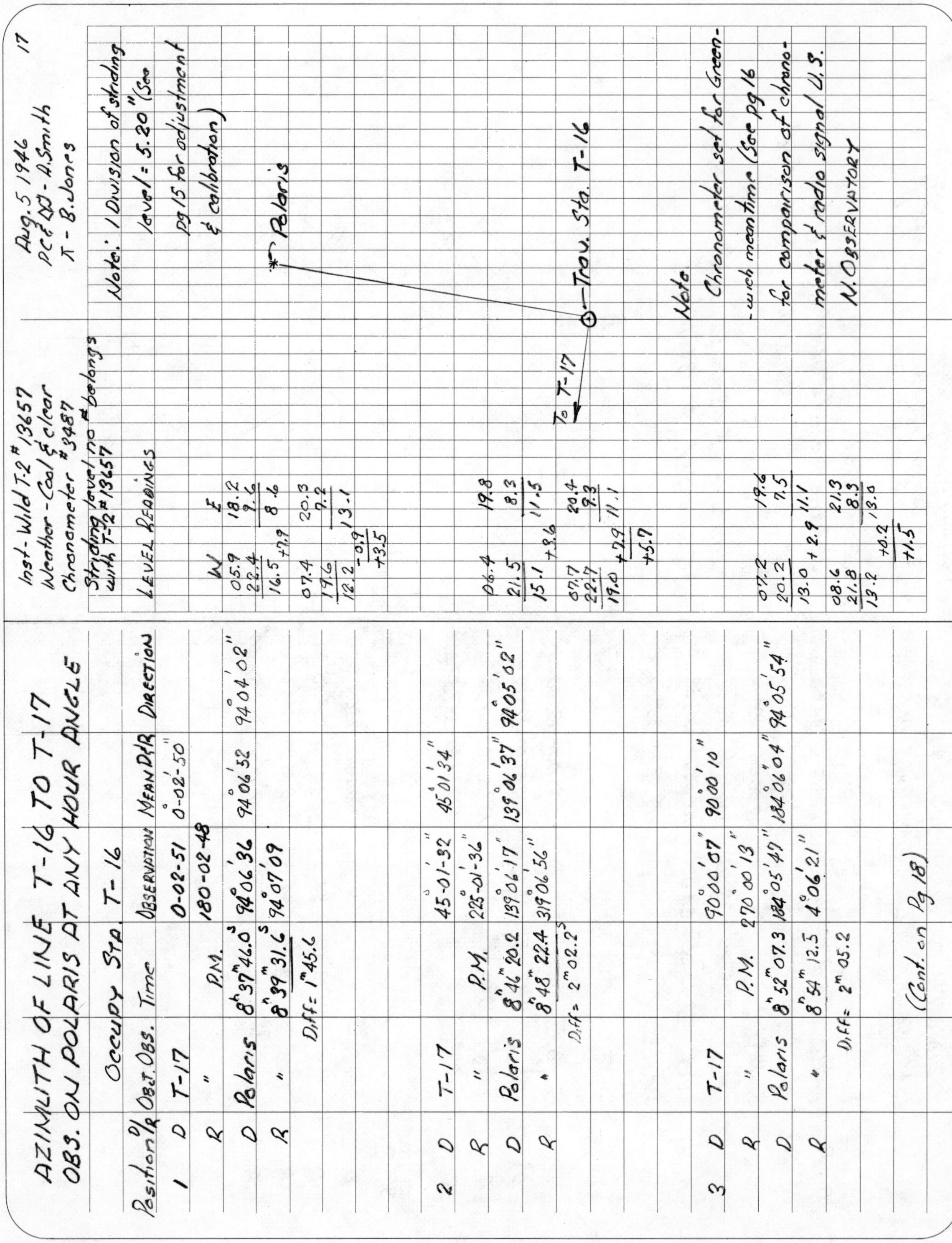

17

AZIMUTH OF LINE T-16 TO T-17
OBS. ON POLARIS AT ANY HOUR ANGLE

Aug. 5 1946
PCE&A - A. Smith
π - B. Jones

Inst. Wild T-2 #13657
Weather - Cool & clear
Chronometer #3487

Note: 1 Division of striding level = 5.20" (See pg 15 for adjustment & calibration)

Striding level no boojangs with T-2 #13657

LEVEL READINGS

Note. Chronometer set for Greenwich meantime (See pg 16 for comparison of chronometer & radio signal U.S. N. OBSERVATORY)

⊙—Trav. Sta. T-16
To T-17
Polaris

Occupy Sta. T-16

Position	D/R	Obj. Obs.	Time	Observation	Mean D&R	Direction
1	D	T-17		0-02-51	0-02-50	
	R	"	P.M.	180-02-48		
	D	Polaris	$8^h37^m46.0$	94°06'36	94°06'52	94°04'02"
	R	"	$8^h39^m31.6$	94°07'09		
			Diff = 1ᵐ45.6			
2	D	T-17		45-01-32	45-01-34	
	R	"	P.M.	225-01-36		
	D	Polaris	$8^h46^m20.2$	139°06'17	139°06'37	94°05'62"
	R	"	$8^h48^m22.4$	319°06'56		
			Diff = 2ᵐ02.2			
3	D	T-17		9°00'07"	9°00'10"	
	R	"	P.M.	270°00'13"		
	D	Polaris	$8^h52^m07.3$	184°05'49"	184°06'04"	94°05'54"
	R	"	$8^h54^m12.5$	4°06'21"		
			Diff = 2ᵐ05.2			

(Cont. on Pg 18)

LEVEL READINGS

W   E
05.9   18.2
22.4   9.6
16.5   8.6
+7.9

07.4   20.3
19.6   9.2
12.2   13.1
      -0.9
      +3.5

06.4   19.8
21.5   8.3
15.1   11.5
+8.6

07.7   20.4
22.7   9.3
19.0   11.1
+2.9   +5.7

07.2   19.6
20.2   7.5
13.0   +2.9 11.1
+0.2

08.6   21.3
21.8   8.3
13.2   13.0
+0.2
+1.5

FIGURE 96

ASTRONOMIC NOTES
POLARIS OBSERVATION
See Section 6-13

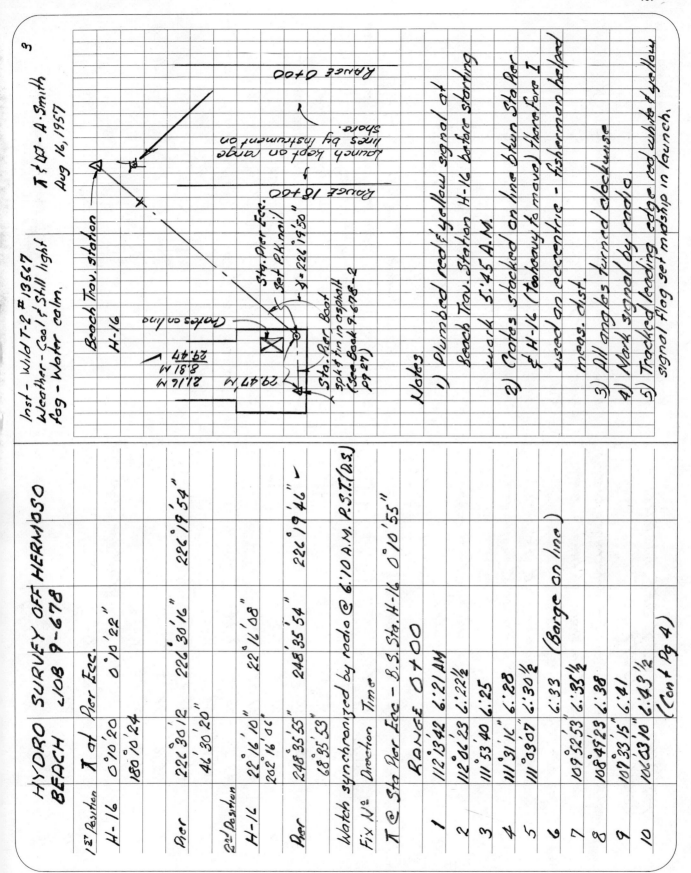

FIGURE 97

HYDROGRAPHIC NOTES
FIXES BY RANGE AND CUTS
See Section 6–14

6

Sea.- Calm, 4' ground swells from N.W. Mar 12, 1956

Weather - Cool N.W Breeze & High Fog.

Party & Equipment operating from 28' Motor Launch "Mary-Gerri."

P.C.S. Boat Plot - A.Smith

Sound Gear - B.White

"A" Sextant & #47 - M.Jones

"B" Sextant - D.Fitz

Helmsman - Capt. C.D. Rexford.

"A" Sextant #12365 - "B" Sextant #12366

Chronometer #3479

Bendix Echo Sounder #B-378934-LN, with transducer head mounted midship on starboard side & 2.1' below water line.

See Pg 5 for calibration of sound gear

See Pg 4 for standardization of chronometer & check of Launch Compass

See Job #13679 Book 1 Pgs 7 to 28 for beach traverse and range line locations & signal colors.

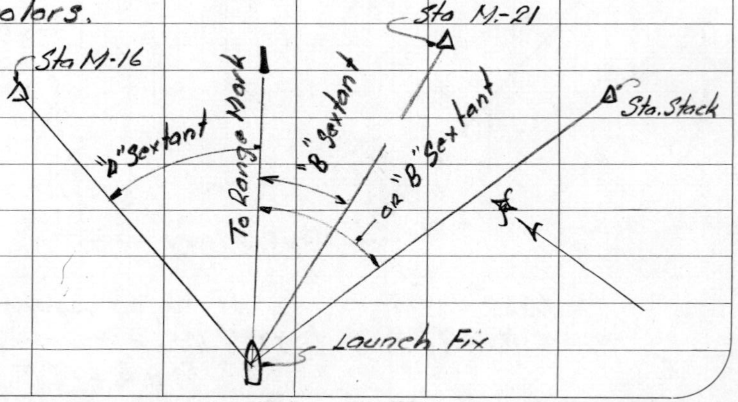

HYDRO. SURVEY OFF MORRO ROCK - JOB 13679

Notes: 1. Range lines run on pre-determined bearing to range signal on shore at launch speed of approx'ly 6 knots.

2. Boat Plot on circular plotting charts

3. Sounding Graph marked & numbered at each "Fix"

4. Sextant angles always from range mark to signal noted.

Fix Nº	"A" Sextant	"B" Sextant	Time	
	LAUNCH ON RANGE 12400			Bearing N 51° 30' E
	M-16	STOCK		
1	42°11'	49°53'	7:32 AM	
2	42°36'	50°20'	7:35 "	
3	42°43'	51°08'	7:38½	
4	43°07'	51°53'	7:41	
5	43°29'	53°08'	7:44	
6	43°57'	54°31'	7:46½	
8	44°26'	55°49'	7:50	
9	45°16'	57°03'	7:53	
10	46°10'	58°37'	7:55½	
		M-21		Can't see Stock
11	47°02'	39°01'	7:59	
12	47°59'	39°32'	8:01½	
13	48°52'	39°58'	8:04	

(Cont on Pg 7)

FIGURE 98

HYDROGRAPHIC NOTES
FIXES USING TWO SEXTANTS
See Section 6-14